D1268811

The Mycetozoans

The Mycetozoans

Lindsay S. Olive

Department of Botany
The University of North Carolina
Chapel Hill, North Carolina

With the Technical Assistance of

Carmen Stoianovitch

Department of Botany
The University of North Carolina
Chapel Hill, North Carolina

Academic Press   *New York   San Francisco   London*   **1975**

A Subsidiary of Harcourt Brace Jovanovich, Publishers

To John Nathaniel Couch

and

To the Memory of William Chambers Coker

ACADEMIC PRESS, INC.
111 Fifth Avenue, New York, New York 10003

United Kingdom Edition published by
ACADEMIC PRESS, INC. (LONDON) LTD.
24/28 Oval Road, London NW1

Library of Congress Cataloging in Publication Data

Olive, Lindsay S
    The mycetozoans.

    Bibliography: p.
    1.   Myxomycetes.   I.   Title.   [DNLM: 1.   Myxomycetes.
QK635.A1 048m]
QK635.A1042   589'.29   74-10200
ISBN 0–12–526250–7

Contents

Preface   ix

Chapter 1   Introduction and Keys to Higher Taxa
          Text   1

MYCETOZOA

EUMYCETOZOA

Chapter 2   Protostelia (Protostelids)
   I. Introduction   11
   II. Occurrence, Isolation, and Maintenance   12
   III. Life Cycle   14
   IV. Classification   18
   V. Descriptions of Taxa   19
   VI. Ultrastructure   38

Chapter 3   Dictyostelia (Dictyostelid Cellular Slime Molds)
   I. Introduction   45
   II. Occurrence, Isolation, and Maintenance   46
   III. Life Cycle   47
   IV. Key to Families and Genera   51
   V. Keys to Species and Descriptions of Taxa   52

    VI. Details of the Life Cycle   67
   VII. Ultrastructure   84
  VIII. Genetics   95

Chapter 4 Myxogastria (Myxomycetes)
     I. Introduction   99
    II. Occurrence, Isolation, and Maintenance in Culture   100
   III. Life Cycle of a Myxomycete   103
   IV. Classification   106
    V. Descriptions of Taxa   109
    VI. Details of the Life Cycle   120
   VII. Ultrastructure   141
  VIII. Genetics   152

ACRASEA

Chapter 5 Acrasea (Acrasid Cellular Slime Molds)
     I. Introduction   159
    II. Occurrence, Isolation, and Maintenance   161
   III. Life Cycle of an Acrasid (*Acrasis rosea*)   162
   IV. Classification   169
    V. Descriptions of Taxa   171
    VI. Ultrastructure   184

ASSOCIATED GROUPS

PLASMODIOPHORINA AND LABYRINTHULINA

Chapter 6 Plasmodiophorina (Plasmodiophorids)
     I. Introduction   193
    II. Occurrence in Nature and Laboratory Maintenance   194
   III. Life Cycle   194
   IV. Classification   200
    V. Ultrastructure   206

Chapter 7 Labyrinthulas (Labyrinthulina and Thraustochytrids)
     I. Introduction   215
    II. Occurrence, Isolation, and Maintenance   216

III. Life Cycle 218
IV. Classification 223
V. Ultrastructure 230

PHYLOGENY

Chapter 8 Phylogenetic Implications
Text 245

Appendix 253
Bibliography 255

Taxonomic Index 283
Subject Index 287

Preface

The purpose of this monograph is to bring together, for the first time in a single volume, comprehensive information on the biology and classification of the mycetozoans and associated groups. There appears to be a need for such a book as a text or reference source in courses on mycetozoans, protozoology, mycology, and developmental biology of lower organisms and as a concentrated source of information for research workers in all aspects of the biology and taxonomy of these organisms. The mycetozoans and their associates remain of prime interest to taxonomists and phylogenists because major new taxa continue to be discovered among them. The latest and probably most primitive subclass (Protostelia) of mycetozoans was described only a few years ago.

In recent years two mycetozoans in particular—*Dictyostelium discoideum* and *Physarum polycephalum*—have become favored laboratory subjects for investigations of development at molecular and ultrastructural levels. Therefore, an attempt has been made to review some of the more important aspects of these investigations as well as the life cycles of other less familiar mycetozoans, some of which have features that make them uniquely suited to similar studies.

Mycetozoans have long intrigued biologists by virtue of their combination within a single life cycle of an animallike feeding stage with a plantlike fruiting stage. Fortunately, many of them can be isolated readily by even a novice and maintained in the laboratory with a minimum of equipment and effort. They are ideal classroom subjects for demonstrating within a short period of time the clear-cut developmental steps that lead to fruiting. Several species are particularly useful for demonstrating protoplasmic flow, cell movement, chemotactically stimulated cell aggregation, and synchronized nuclear division. This book informs the reader of where to find mycetozoans, of how to isolate and cul-

ture them, of their life cycles and ultrastructure, and of some of the experiments that may be performed with them. Undoubtedly, new research
problems will suggest themselves to the individual reader in relation to
his own particular interests.

Appreciation is extended to the numerous investigators who have contributed many of the illustrations and to those who have made construc-
tive suggestions following critical reviews of several chapters; namely,
Dr. John T. Bonner (Chapter 3), Dr. William D. Gray (Chapter 4),
Dr. Frank O. Perkins (Chapter 7), and Dr. Kenneth B. Raper (Chapters
1, 3, and 5). The author especially acknowledges the constant and in-
valuable assistance of Miss Carmen Stoianovitch, who discovered and
isolated many of the new mycetozoans described in our laboratory over
the past fifteen years and who has made important contributions to the
preparation of this book. The assistance of Mrs. Lindsay S. Olive in
preparation of the manuscript and of Miss Marion Seiler in preparing
most of the drawings is also gratefully acknowledged. The many years
of support of our research on mycetozoans by the National Science Foun-
dation as well as grants-in-aid from the Brown-Hazen Fund of the Research Corporation are greatly appreciated.

Lindsay S. Olive

CHAPTER 1

Introduction and Keys to Higher Taxa

The mycetozoans and their associates have long intrigued zoologists, botanists, and phylogenists interested in lower organisms. The combination of animal-like feeding stage and plant-like fruiting body within the same life cycle has resulted in their being classified in both the animal and plant kingdoms. Recent discoveries relating to their life cycles, biology, and taxonomy have introduced new concepts concerning their interrelationships and their possible evolutionary affinities with simpler forms.

Lately, several species of mycetozoans, especially the cellular slime mold *Dictyostelium discoideum* and the myxomycete *Physarum polycephalum*, have become widely used experimental subjects in the study of morphogenesis. The particular advantages of these organisms for this type of research are related to their ease of culture in the laboratory and their relatively rapid passage through a life cycle consisting of a sequence of distinct stages. An increasing amount of information is now accumulating that links these developmental stages with analyzable biochemical changes in the protoplasts. The end result is certain to be a better understanding of cell processes controlling morphogenesis. Although this subject is discussed in greater detail in the appropriate chapters, one example of the applicability of such studies to higher organisms may be noted here. Both human leukocytes and the amoebae of *Dictyostelium* show remarkable similarities in general appearance, mode of movement, presence of the conventional eukaryotic cell organelles, and positive tactile response to the nucleotide cyclic 3′,5′-adenosine monophosphate (cyclic AMP), both types of cells moving toward the attractant, whether it be in pure form or diffusing from the cells of bacteria.

Systems of classifying mycetozoans and associated groups have varied greatly, depending upon the outlook of the classifier. The outstanding mycologist De Bary (1859, 1887) erected the class Mycetozoa for the

Myxomycetes (plasmodial slime molds) and the Acrasieae (cellular slime molds) of van Tieghem (1880), which he recognized as coordinate orders. *Plasmodiophora* and *Labyrinthula* were treated in a section on "Doubtful Mycetozoa." De Bary considered the Mycetozoa outside the bounds of the plant kingdom. He also believed that they had evolved independently of the fungi, a group in which a number of mycologists subsequently placed them. Haeckel (1868) considered mycetozoans as neither plants nor animals but as primitive forms that had not yet evolved into members of either of the two major kingdoms. This led him to erect the new kingdom Protista in which he placed, among others, typical protozoans, mycetozoans, labyrinthulids, and fungi. Lankester (1885) placed the class Mycetozoa, along with the coordinate class Proteomyxa, in grade A—the Gymnomyxa—of the phylum Protozoa.

More recently, Copeland (1956) proposed a drastically different four-kingdom system of classification that, while controversial to say the least, stimulated renewed thought on the subject. He included the class Mycetozoa, consisting of myxomycetes and plasmodiophorids, and the class Sarkodina, containing cellular slime molds and labyrinthulids, along with other classes of organisms commonly recognized as protozoans, in the phylum Protoplasta of the kingdom Protoctista. The algae and fungi were included in the same kingdom. By contrast, Honigberg *et al.* (1964) chose a more conventional approach to the classification of the protozoa. They placed the subclasses Mycetozoia (cellular slime molds, myxomycetes, plasmodiophorids) and Labyrinthulia (labyrinthulans) in class Rhizopodea, superclass Sarcodina, and subphylum Sarcomastigophora of the phylum Protozoa. In proposing a revised classification of the Mycetozoa, Olive (1970) tentatively treated the group as a class within the system of Honigberg *et al.* but suggested that a new taxonomic concept published by Whittaker (1969) might eventually be found preferable. Four subclasses were recognized by Olive: Protostelia (protostelids), Dictyostelia (dictyostelid cellular slime molds), Acrasia (acrasid cellular slime molds), and Myxogastria (myxomycetes). For the first time, the dictyostelid and acrasid cellular slime molds were separated into two major groups on the basis of contrasting types of amoebae and modes of sorogenesis. The protostelids, discovered in our laboratory, are the simplest and probably the most primitive of the mycetozoans; therefore, their phylogenetic implications are of particular interest.

In his classification of organisms Whittaker (1969) has adopted many of Copeland's proposals, accepting with certain revisions his four kingdoms and erecting a new one for the fungi. This five-kingdom system has been accepted by at least one modern biology text (Raven and

Curtis, 1971) and, with modifications, is the one that we have found most acceptable. The five kingdoms recognized by Whittaker are Monera (bacteria and blue-green algae), Protista (essentially the protozoans and simple algal forms), Fungi, Plantae (multicellular plants), and Animalia (multicellular animals). The separation of the prokaryotic Monera from the eukaryotes hardly needs further justification in the light of modern knowledge of their ultrastructure and biochemistry. Moreover, the fungi stand apart from other groups in several significant ways, including their typically mycelial trophic stage with nonphotosynthetic absorptive nutrition. As in any system of classification, difficulties arise in efforts to draw sharp lines between groups, and in Whittaker's proposal such problems are found in the apportioning of algal forms among the Protista and Plantae. On the whole, however, we find it to be a satisfactory system that has fewer pitfalls for students than have more conventional ones. Whittaker's classification of the kingdom Fungi, in which he includes the mycetozoans, is as follows.

Subkingdom Gymnomycota
    Phylum Myxomycota (myxomycetes)
    Phylum Acrasiomycota (cellular slime molds)
    Phylum Labyrinthulomycota (labyrinthulids)
Subkingdom Dimastigomycota
    Phylum Oomycota (oomycetes)
Subkingdom Eumycota
  Branch Opisthomastigomycota
    Phylum Chytridiomycota (chytridiaceous fungi)
  Branch Amastigomycota
    Phylum Zygomycota (zygomycetes)
    Phylum Ascomycota (ascomycetes)
    Phylum Basidiomycota (basidiomycetes)

Whittaker placed the phyla Hyphochytridiomycota (hyphochytrids) and Plasmodiophoromycota (plasmodiophorids) in the kingdom Protista. The phylum Deuteromycota (Fungi Imperfecti) may be added to the Amastigomycota.

Olive (1969), in commenting on Whittaker's classification, recommended that the Gymnomycota be transferred from the fungi to the kingdom Protista on the basis that they have holozoic rather than absorptive nutrition (except for the labyrinthulids) and show little evidence of being related to the fungi. It was also suggested that the plasmodiophorids (and proteomyxids), intracellular parasites with small plasmodia, be included in the Gymnomycota, and that the anteriorly uniflagellate Hyphochytridiomycota, which have absorptive nutrition and may have evolved from oomycetous forms through loss of the whiplash flagellum, be transferred from the protists back to the fungi (where they have generally been classified by mycologists). These recommendations are incorporated into the revised classification of fungi and mycetozoans (with associated groups) that we propose below. A number of Whittaker's taxa are accepted at a reduced rank. In addition, we believe that the distribution of the two major pathways of lysine biosynthesis— α-ϵ-diaminopimelic acid (DAP) and α-aminoadipic acid (AAA) pathways—as determined by Vogel (1964) in the fungi should be taken into account in any current treatment of this group, and they are therefore indicated in the following classification of the fungi.

Kingdom Fungi

Phylum Pantonemomycota (planonts with anteriorly directed mastigonemate flagellum and posteriorly directed whiplash flagellum, or with anterior flagellum only; cellulose typically present in cell wall; DAP lysine pathway)
    Class Oomycetes (planonts with anteriorly and posteriorly directed flagella; mostly mycelial forms)
    Class Hyphochytridiomycetes (planonts with anteriorly directed mastigonemate flagellum only; typically chytrid-like; cellulose present or absent)

Phylum Eumycota (planonts with single posterior whiplash flagellum, or motile stages absent; cellulose generally absent from cell walls; AAA lysine pathway)
    Subphylum Opisthomastigomycota (planonts present; mostly aquatic forms)
    Class Chytridiomycetes (chytridiaceous fungi)
    Subphylum Amastigomycota (planonts lacking; typically terrestrial and mycelial)
    Class Zygomycetes (conjugation fungi)
    Class Ascomycetes (sac fungi)
    Class Basidiomycetes (club fungi)
    Class Deuteromycetes (imperfect fungi)

Our proposed classification of the mycetozoans and their associates is as follows.

Kingdom Protista

Phylum Euprotista (most protozoans and simple algal forms)
Phylum Gymnomyxa (mycetozoans and associates). Trophic stage typically amoeboid and holozoic (but primarily absorptive in associated groups); multinucleate plasmodia or multicellular pseudoplasmodia present or absent; typically producing aerial fruiting bodies or, in associated groups, zoosporangia and/or sori of resting spores; syngamy present or absent
    Subphylum Mycetozoa (mycetozoans). Producing aerial fruiting bodies; flagella, when present, in an unequal, less often equal apical pair and nonmastigonemate
    Class Eumycetozoa (true mycetozoans). Amoeboid stage producing filose pseudopodia; stalk tube typically present in fruiting bodies of the first two classes and in some myxomycetes

Subclass Protostelia (the protostelids). Trophic stage varying from simple amoebae to plasmodia that lack shuttle movement of protoplasm; flagellate cells present or absent; fruiting bodies consisting of one to several spores on a narrow hollow stalk; sexual reproduction known in only one species.

Subclass Dictyostelia (dictyostelid cellular slime molds). Nonflagellate; amoebae aggregating to form a multicellular pseudoplasmodium that gives rise to a multispored fruiting body; stalk tube present; occurrence of sexual reproduction indicated in some species

Subclass Myxogastria (myxomycetes or true slime molds). Flagellate cells present; major trophic stage a multinucleate coenocytic plasmodium that typically shows shuttle movement of protoplasm; fruiting bodies multi-spored; syngamy and meiosis present in life cycle

Class Acrasea (acrasid cellular slime molds). Trophic stage comprised of amoeboid cells with lobose pseudopodia; flagellate cells known in only one species; amoeboid cells aggregating to form a pseudoplasmodium that gives rise to a fruiting body lacking a stalk tube; sexual reproduction unknown

Subphylum Plasmodiophorina. Obligate intracellular parasites with minute plasmodia; zoospores produced in zoosporangia and bearing an anterior pair of unequal, nonmastigonemate flagella; resting spores formed in compact sori or loose clusters within the host cells; sexual reproduction reported in some

Class Plasmodiophorea (the plasmodiophorids). With characteristics of the subphylum

Subphylum Labyrinthulina. Trophic stage producing an ectoplasmic net that functions in absorptive nutrition and in *Labyrinthula* provides channels for cell movement; aquatic forms with zoosporangia that produce planonts with anteriorly directed mastigonemate flagellum and posteriorly directed nonmastigonemate flagellum; evidence of sexual reproduction found in only one isolate of *Labyrinthula*.

Class Labyrinthulea (labyrinthulas and thraustochytrids). With characteristics of the subphylum

Whittaker's taxon Gymnomycota has been replaced by the much earlier one, Gymnomyxa, of Lankester (1885), modified to exclude the relatively simple forms that lack specialized fruiting structures such as soro-carps, cystosori, and zoosporangia (a heterogeneous group classified by Lankester among the proteomyxans). Because we do not consider the Gymnomyxa to be fungi, substitutions for the -*mycota* suffixes of other Whittaker taxa have been made. The concept of the subphylum Myce-tozoa adopted here is essentially that of De Bary, who considered the group to be comprised of organisms with amoeboid, holozoic trophic stage and aerial fruiting bodies. Subphyla Plasmodiophorina and Laby-rinthulina are included primarily as a matter of convenience, because they have frequently been classified or tentatively allied with the myce-tozoans and their actual relationships are obscure. We consider neither subphylum to be related to the Mycetozoa. Ainsworth (1973) divides the Fungi into Myxomycota and Eumycota. His key to the Myxomycota contains four "classes": Acrasiomycetes, Labyrinthulales, Myxomycetes, and Plasmodiophoromycetes. He then excludes the Labyrinthulales from the subdivision. The Plasmodiophoromycetes have been placed by Waterhouse (1973) in subdivision Mastigomycotina of the Eumycota, in which are also included classes Zygomycotina, Ascomycotina, Basidiomycotina, and Deuteromycotina.

The discovery in our laboratory (see Olive, 1970) of the primitive and relatively simple Protostelia has indicated a closer relationship between myxomycetes and dictyostelid cellular slime molds than previously realized. Both groups appear to have evolved, albeit along separate lines, from the protostelids. At the same time, accumulating evidence increasingly alienates the acrasid cellular slime molds (Acrasea) from other mycetozoans. We are inclined to agree with the growing bulk of opinion among protozoologists (e.g., Bovee and Jahn, 1965) that types of amoebae and amoeboid movement should be given considerable weight in the classification of these lower forms. As discussed further in the appropriate chapters, the dictyostelid amoeba, with its filose pseudopodia, is quite different in appearance and mode of locomotion from the acrasid amoeba, with its eruptive lobose pseudopodia. The acrasid fruiting body is also sharply distinguished from that of the dictyostelids in lacking a cellulosic stalk tube.

These and other differences to be discussed suggest the independent origin of these two groups of cellular slime molds. Earlier investigators have been inclined to associate them more closely on the basis that all exhibit cell aggregation leading to pseudoplasmodial development and fruiting. However, it has long been known that a similar phenomenon occurs in the unrelated myxobacteria, where it must have evolved independently. Raper and Alexopoulos (1973) have described aggregation of myxamoebae to form columns of microcysts in an aberrant isolate of the myxomycete *Didymium nigripes*. We have recently discovered, on dead plant materials collected in both North Carolina and France, a remarkable ciliate that forms aggregates of cells which, behaving in amoeboid fashion, secrete a common sheath around themselves and rise from the substrate to form aerial fruiting bodies (Figs. 1 and 2). The sheath becomes wrinkled and desiccated beneath the rising cell mass to form a hollow tapered stalk, and the cells become encysted in a round terminal sorus when stalk elongation ceases. Each germinating cyst gives rise to a single ciliate cell. Unfortunately, we have been unable to culture this organism and to study its development in greater detail. However, its discovery further demonstrates that cell aggregation leading to fruiting has evolved on more than one occasion and should therefore not be overemphasized in the classification of mycetozoans.

We have chosen to omit the genus *Sappinia* from the mycetozoans.

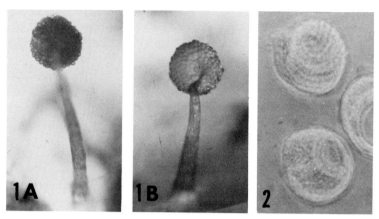

Figs. 1 and 2. Unidentified ciliate. Fig. 1, A and B. Sorocarps formed from aggregates of cells. Each has a hollow stalk and an apical sorus of encysted cells. ×36. Fig. 2. Germinating cysts, each containing a single ciliate cell. ×2000.

E. W. Olive (1902), with some misgivings, tentatively included it among the cellular slime molds. It is characterized by comparatively large amoebae that form rather casual aggregates of encysted cells as well as upright, club-shaped resting cells on such substrates as dung and dead vegetation. The latter cells are surrounded by a protective sheath but have no cell wall, and when pushed back onto a moist substrate they revert almost at once to the amoeboid state.

In contemplating the evolution of terrestrial organisms from aquatic protists, it is common to think only of the two major phylogenetic lines, one leading to the higher animals and one to the higher plants. The Mycetozoa, however, represent a group in which several independent evolutionary lines have probably led from aquatic protists to terrestrial forms with aerial fruiting bodies (Fig. 251). The major distinction is that none of these forms appears to have evolved in complexity beyond the group (Protista) in which they originated. Discussions of the various developmental systems that have evolved in these different phylogenetic lines, insofar as they are known to us, are integral parts of subsequent chapters. At the same time, it becomes apparent that there is much need for further research on the biology, development, and classification of the mycetozoans.

Mycetozoa

Eumycetozoa

CHAPTER 2

Protostelia (Protostelids)

I. INTRODUCTION

The discovery that led to the establishment of the new mycetozoan subclass Protostelia was incidental to an attempt to isolate a new cellular slime mold, *Acrasis rosea* Olive & Stoian., from dead grass inflorescences. In addition to the cellular slime mold, an organism with minute fruiting bodies, each consisting of a slender tubular stalk bearing a single terminal spore, was isolated. The latter was named *Protostelium mycophaga* Olive & Stoian. (1960). Both organisms had trophic stages consisting of uninucleate orange-pigmented amoebae that were found to ingest cells of the pink yeast *Rhodotorula mucilaginosa* isolated from the same substrate. However, the amoebae of *Protostelium*, with their filose pseudopodia, were readily distinguishable from those of the *Acrasis*, with lobose pseudopodia. *Protostelium mycophaga* proved to be only one of a varied and widespread group of organisms that are probably the simplest known mycetozoans.

Although the new organism was first placed in the same family with the dictyostelid cellular slime mold genus *Acytostelium*, which has similar tubular stalks, it conspicuously differed from that genus in lacking an aggregation stage leading to fruiting. As new protostelids were discovered, a new order and finally a new subclass were established to accommodate them (Olive, 1970). At the present time, three families, nine genera, and 17 species of protostelids have been described. We have either isolated or observed on their natural substrates numerous other species that have not yet been described.

As will become apparent in the discussions that follow, the protostelids are of special interest from an evolutionary standpoint because they appear to occupy a primitive position in the Mycetozoa, and they show evidence of having evolved from simple algal protists and of having given rise to both dictyostelid cellular slime molds and myxomycetes.

II. OCCURRENCE, ISOLATION, AND MAINTENANCE

It may seem strange that a group as prolific and widespread as the protostelids has (with the exception of *Ceratiomyxa*) gone unnoticed until recently. However, in isolation plates the microscopic sporocarps are easily mistaken for sporulating structures of certain mucoraceous or imperfect fungi and on any except the weakest agars are likely to be obscured by fungal overgrowth. On weakly nutrient agars, such as oak bark, hay infusion, or weak lactose–yeast extract agar (see Appendix), they may be readily isolated from a great variety of substrates, such as moribund capsules, pods, fleshy fruits, and flowers, as well as from bark, rotting wood, soil, and dung. The majority are found on dead attached plant parts rather than in soil, humus, or dung, probably because they are unable to cope with antagonistic microorganisms in the latter substrates. They are widely distributed, and a number of species have been found in various parts of the world. Although species of *Ceratiomyxa* are readily visible in nature on rotting wood, and although on rare occasions the fruiting bodies of one or two other species may be so abundant that they appear as a whitish bloom on the substrate, protostelids are generally too inconspicuous to be seen in the field. Their potential substrates must be returned to the laboratory and plated out. Soil samples may be suspended in water and plated onto the agar medium, but another method is preferable for other types of substrates. The latter are first soaked in water for a half hour or so, removed and blotted, and then broken into small pieces that are arranged in a concentric fashion around the culture plate, about six inocula per plate. If protostelids are present, they may begin to appear within 3 days either on the substrate or on the nearby agar surface. However, plates should not be discarded too early because some protostelids develop more slowly and do not appear until about a week after plating.

Bark and rotting wood may also be thoroughly moistened and placed in a petri dish with a thin layer of water on the bottom. After a few days the substrates should be examined under the dissecting microscope. In addition to protostelids, some minute and frequently overlooked myxomycetes, as well as cellular slime molds, may be discovered by this method. A number of these mycetozoans seem to occur only on bark, which has its own unique microenvironment. For some unknown reason, quite a few of the mycetozoans found on bark are very resistant to culture in the laboratory. Even some with fruiting bodies indistinguishable from those of well-recognized, easily cultured species isolated from other substrates have failed to grow in laboratory culture. It is clear that we must learn much more about bark as a natural substrate before these problems can be solved.

On agar media the sporocarps of protostelids are readily distinguished from the fruiting structures of fungi in that they do not arise from hyphae. Of course, when the spores are transferred to fresh media they germinate to produce amoeboid or flagellate cells rather than germ tubes. Spores may be readily transferred by touching them with a needle to which a small bit of agar has been attached. The agar causes them to adhere better and protects them from drying in the process. Protostelids with deciduous spores may be collected on a fresh plate of medium by inverting the original isolation plate over it.

If the protostelid is unknown, several different food organisms should be tested to find out which are most suitable for its growth. Some of the most useful food organisms have proved to be *Flavobacterium* sp., *Escherichia coli, Aerobacter aeruginosa, Rhodotorula mucilaginosa,* and especially several unidentified bacteria and yeasts that have been isolated from the natural substrates. It is also usually necessary to experiment with the pH of the medium, as some species have proved to be very demanding in this respect. In general, protostelids will grow between pH 5 and 8.

A number of species are easily isolated and grown in culture. For example, the spores of *Protostelium mycophaga* germinate readily and the amoeboid protoplasts soon begin feeding on yeast cells and dividing to produce a large population of orange-pigmented amoebae, with sporulation occurring in about 3 days. Other species require an extended period of adaptation on the agar medium, in the presence of the food organism, during which growth is very slow. Eventually, the protoplasts show increased activity, undergoing growth and division followed by sporulation, the latter sometimes after more than a month in culture. Once the period of adaptation is over, there is much less of a lag when fresh transfers of spores or amoebae are made. Occasionally, an isolate grows but fails to fruit in culture, whereas the spores of others fail to germinate on any of the media tested or, after germination, their protoplasts do not survive on the food organisms offered.

Once a protostelid has been isolated on a suitable medium and food organism and induced to sporulate, there is usually no problem in maintaining it and inducing it to fruit later. Cultures are best maintained in test tubes or plates at 17°C and transferred to fresh media about once a month. In the case of a number of protostelids, perhaps most, cultures containing spores and cysts may be dried down and revived after 6 months of storage at 17°C.

III. LIFE CYCLE

Since there are many variations in the life cycles of protostelids, only a very generalized outline is given at this point and some individual life cycles are discussed under the generic descriptions (Figs. 25 and 34). The spore germinates to produce an amoeboid or flagellate protoplast that feeds on cells of the food organism (Fig. 4) and undergoes division (Fig. 5), with some of the resulting protoplasts giving rise to sporocarps. Protoplasts of the trophic stage (Figs. 3 and 4) contain one or more contractile vacuoles, food vacuoles, and one to many nuclei. The nucleus typically contains a single centrally located nucleolus, but just after division each daughter nucleus is seen to contain several small pronucleoli that eventually fuse into one, although completion of the process is sometimes delayed (Fig 3). The nuclei often show striking variations in volume, probably indicating different ploidy levels. Anastomosis of protoplasts with the establishment of cytoplasmic bridges has also been observed and offers a means by which cytoplasmic exchanges may occur between cells. Protoplasts with more than one nucleus have often been observed to pull out into two major parts, with one or more nuclei in each and with an ever narrowing strand connecting them. The strand finally becomes so attenuated that it snaps in two, and two

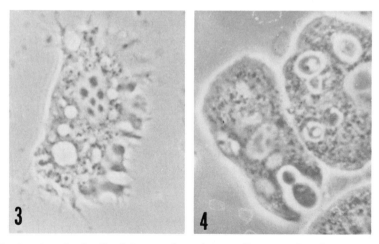

Fig. 3. Amoeboid cell of *P. irregularis* showing filose pseudopodia. The pronucleoli, following mitotic cell division, have not yet fused into a single nucleolus. Phase contrast ×1200.

Fig. 4. Amoeboid cell of *P. mycophaga* ingesting a budding yeast cell. Phase contrast ×1900. From Olive (1967).

independent cells are formed. This type of division is called "plasmotomy" (Fig. 6). It should not be confused with mitotic cell division,
during which division of the protoplast is synchronized with post-
telophase of nuclear division and the two daughter cells, equal in size,
are separated by a medial constriction of the mother cell (Fig. 5).

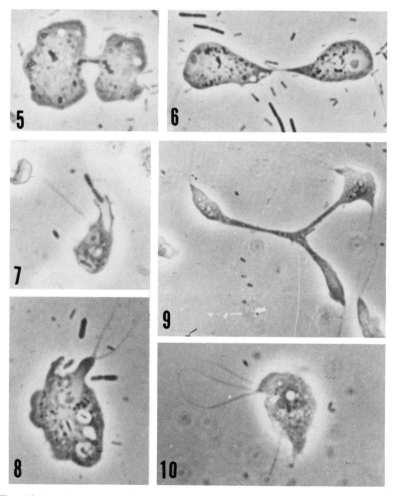

Figs. 5–10. *Cavostelium bisporum,* phase contrast. Fig. 5. Mitotic cell division,
×2500. Fig. 6. Plasmotomy: each of the two separating parts has a nucleus and
a flagellum, ×1850. Figs. 7 and 8. Flagellate cells ingesting bacteria, ×1850. Fig. 9.
Protoplast with several flagella (and nuclei) being pulled in various directions
by uncoordinated flagellar activity, ×1250. Fig. 10. Cell with four flagella, ×1850.
Figs. 5 and 6 from Olive (1967); Figs. 7–10 from Olive and Stoianovitch (1966c).

Flagellate cells typically have a single long flagellum (Fig. 7), with or without a short flagellum at its base, or less often two long flagella (Fig. 8), and supernumerary flagella are not uncommon (Figs. 9 and 10). When the flagellate cell divides mitotically, the flagella first disappear, then reappear shortly after the cells separate. Even cells with supernumerary flagella lose all of them before mitotic division. There is no loss of flagella from flagellate cells undergoing plasmotomy (Fig. 6). At least in some flagellate forms, the single long flagellum has a short reflexed one at its base. All flagella lack mastigonemes. Structures called "pseudoflagella" often arise successively from the flagellate tip of the cell and move back to the posterior end, where they disappear. They appear to be peculiar filose pseudopodia that are unusually long and slender. Filose pseudopodia are characteristic of protostelids (Fig. 3). They are temporary, narrow hyaloplasmic extensions of the cell that may aid in food entrapment, in cell movement, in extending the reticulations of a plasmodial protoplast, or even in forming rhizoidlike extensions of the protoplast into the substrate. The forward side of a migrating cell is hyaloplasmic and, like the pseudopodia, lacks major cell organelles.

Single amoebae or segments of plasmodial protoplasts may become encysted on the substrate and enter a resting period. Encystment generally increases with age of the culture and probably with changes in pH, accumulation of staling products, and depletion of food. These cysts usually germinate to produce the trophic stage when transferred to fresh medium.

In protostelids with plasmodia, sporulation is preceded by segmentation of the plasmodium into uninucleate or plurinucleate prespore cells, whereas in forms characterized by a uninucleate trophic stage any amoeba may function as a prespore cell. The process of sporulation is diagramatically illustrated in Fig. 11. The prespore cell (A) becomes hemispherical and then hat-shaped (B) as the protoplasm becomes concentrated in the central area. A thin, protective sheath is deposited over the prespore cell at this time. It covers the entire sporocarp throughout its development, probably protecting the sporogen from desiccation. It is probably homologous with the slime sheath of dictyostelids and myxomycetes. The hat-shaped stage is followed by a stage (C) during which the protoplast withdraws from the brim of the hat, allowing the sheath in this area to collapse on the substrate and become the supporting disc of the sporocarp. At the same time a finely granular hemispherical area of cytoplasm differentiates in the basal region of the prespore cell. This is the steliogen, which now begins to secrete the stalk tube (D). As the protoplast rises at the tip of the elongating stalk tube, it maintains a narrow extension into the upper end of the tube, which probably explains why a hollow tube instead of a solid stalk filament is formed (E). When the stalk reaches its full length, the protoplast at the tip secretes a wall around itself and becomes a spore. The mature sporocarp consists of a slender stalk tube bearing a spore, the whole bounded externally by a sheath that expands below into a basal disc (F). In apophysate protostelids, such as the one illustrated here, the apophysis is thought to be produced by the steliogen just before encystment of the protoplast to form the spore.

In many, if not most, protostelids the entire process of stalk formation is completed within 30–60 minutes. In most species the spores are immediately germinable, but in at least one undescribed species with reticulate-walled spores resembling those of some myxomycetes, a resting period is required. A number of protostelids have deciduous spores and one unusual species (*Schizoplasmodium cavostelioides*) produces ballisto-spores that are forcibly expelled from their stalks. Deciduous-spored species occur mostly on dead attached plant parts, and their spores are most likely dispersed primarily by wind and air currents. A number of other species have nondeciduous spores that are probably distributed chiefly by mites and small insects.

When sporulating cultures of protostelids are flooded with water, the sporocarps float upright on the surface of the water, with only their basal discs in contact with the water surface. Indeed, they are not easily upset from their upright position.

Sexual reproduction has been demonstrated only in *Ceratiomyxa fruti-*

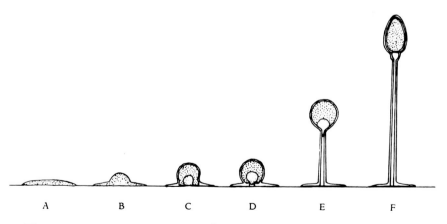

Fig. 11. Sporogenesis in *Nematostelium ovatum:* (A) prespore cell; (B) hat-shaped stage; (C) appearance of steliogen; (D) beginning of stalk formation; (E) later stage in stalk development (note extension of steliogen into stalk tube); (F) mature sporocarp. From Olive and Stoianovitch (1966b).

culosa, in which it takes the form of syngamy between similar flagellate cells and meiotic divisions in the spores. There are a few other protostelids in which a sexual process seems not unlikely but remains undemonstrated. On the other hand, most protostelids appear to have no place in their life cycles for plasmogamy and meiosis.

IV. CLASSIFICATION

The subclass Protostelia and the single order Protosteliida are characterized by having a holozoic trophic stage that varies from uninucleate amoeboid or amoeboflagellate cells to multinucleate reticulate plasmodia. When plasmodia are present they differ from those of myxomycetes in lacking the rhythmic reversible flow of protoplasm but they do show a certain degree of synchrony in nuclear division. Following mitosis in protostelids, the nucleolus, which has disappeared during the division, reappears in each daughter nucleus as several small pronucleoli that fuse together to form a single central nucleolus.

The group is further characterized by the production of filose pseudopodia, although lobose ones may also be formed. The movement of these protoplasts, however, is relatively slow and unlike that of the more rapidly moving amoebae of acrasid cellular slime molds, which have only lobose pseudopodia. Flagellate cells are present in two of the families but are absent in the third.

The sporocarp typically consists of a slender tubular stalk bearing a single spore at its tip, although some species may have two or even four (rarely more) spores. Although the spores are generally smooth walled, in a few protostelids they show characteristic external markings. The spores have a variety of shapes, being round, ovate, pyriform, or peanut-shaped. At germination they produce either amoeboid protoplasts or flagellate cells characteristic of the species.

The presence or absence of flagella has often been used as a major criterion in separating groups of organisms in systems of classification. Because of its convenience, this character has also been used in the classification of protostelids; a single family of nonflagellate forms has been separated from two families with flagellate cells primarily on this basis. Its validity for distinguishing actual relationships in this group must be questioned, however. For example, *Protostelium mycophaga* and the rare *Planoprotostelium aurantium* are very much alike in most respects, including their orange pigmentation, and are distinguished primarily on the basis of flagellation of the trophic stage in the latter and its absence in the other. It is most likely that *Protostelium myco-*

phaga evolved from *Planoprotostelium aurantium* by loss of flagellation and that other nonflagellate forms have arisen in similar fashion and on several occasions from flagellate protostelids. It is thought that one of these nonflagellate lines may have given rise to the dictyostelid cellular slime molds (Fig. 251).

Stalks and walls of spores and cysts often turn bluish in chloriodide of zinc, and stalks of one of the larger protostelids tested with iodine–potassium iodide and concentrated sulfuric acid (M. Dykstra, unpublished data) become lavender in color, strongly indicating the presence of cellulose in these structures.

Key to the Families and Genera

Flagellate cells present in the life cycle
  Sporocarps produced directly on substrate . . . .   Cavosteliidae
    Trophic stage comprised of uninucleate to plurinucleate amoeba-like protoplasts
      Sporocarps with relatively short stalks, 1- to 2-spored. . .   *Cavostelium*
      Sporocarps with elongate stalks
        Sporocarps like those of *Protostelium mycophaga*, mostly 1-spored . . . .   *Planoprotostelium*
        Sporocarps with long flexuous stalks, typically 2- to 4-spored . . . .   *Protosporangium*
    Trophic stage at maturity a reticulate plasmodium . . . . . .   *Ceratiomyxella*
  Sporocarps produced on upright gelatinous (later firm) extensions of the plasmodium . . . .   Ceratiomyxidae
                                Single genus: *Ceratiomyxa*
Flagellate cells absent . . . .   Protosteliidae
  Spores typically deciduous; conspicuous to reduced subspore apophysis present
    Trophic stage a reticulate plasmodium
      Spores forcibly discharged from short stalks . . . .   *Schizoplasmodium*
      Spores deciduous but not forcibly discharged, borne on long stalks . . . .   *Nematostelium*
    Trophic stage comprised of uninucleate to plurinucleate amoeboid protoplasts . . . .   *Protostelium*
  Spores nondeciduous and nonapophysate
    Trophic stage a reticulate plasmodium . . . .   *Schizoplasmodiopsis*
    Trophic stage comprised of uninucleate amoeboid protoplasts . . . .   *Protosteliopsis*

V. DESCRIPTIONS OF TAXA

A. Cavosteliidae L. Olive (1964a)

This family contains those protostelids that have flagellate cells at some stage in the life cycle, a trophic stage ranging from uninucleate amoeboid or amoeboflagellate cells to multinucleate reticulate plasmodia, and sporocarps that develop directly on the substrate. The sporocarp produces from one to four (rarely more) spores. Certain members of the group show striking similarities to the simplest known myxomycetes, a probable indication of ancestral relationships. Four genera have been described: *Cavostelium, Ceratiomyxella, Planoprotostelium, and Protosporangium.*

Cavostelium L. Olive (1964a) is a genus of two known species with short-stalked sporocarps bearing one or two spores and a trophic stage consisting primarily of uninucleate amoeboid cells but also including some multinucleate ones that might be considered plasmodial, although they remain nonreticulate. Encysted protoplasts are variable in size and shape (Fig. 18). The flagellate cells are uninucleate, with a single long anterior flagellum and a short reflexed flagellum that is difficult to see, or there may be two long flagella (Figs. 7 and 8). These flagella are associated with the nucleus, and when they change position on the cell the nucleus follows. Supernumerary flagella are also frequently present, although independent of the nucleus (Fig. 10). They tend to pull the cell in various directions (Fig. 9), often stretching parts of the cell into narrow extensions that, on occasion, may even become detached from the cell. These detached enucleate cell fragments may continue to show flagellar activity for an hour or so. The flagellate cells of *Cavostelium,* like those of most other flagellate forms, are rather poor swimmers and move irregularly and slowly. They never seem to lose their amoeboid nature or the ability to ingest food (Figs. 7 and 8).

The minute fruiting bodies of *Cavostelium* are among the smallest produced by protostelids (Fig. 18). However, their simple appearance may be deceptive in view of ultrastructural studies that show an advanced spore wall structure similar to that in primitive myxomycetes. Under the light microscope the spore wall appears to be minutely punctate (Figs. 23 and 24) but thin sections under the electron microscope reveal these dots to be the tips of structures that pass through the entire wall (Figs. 69 and 70) as in the myxomycete genus *Echinostelium* (Figs. 161–163.

Cavostelium apophysatum L. Olive (1964a), worldwide in distribution on moribund attached plant structures, has 1- to 2-spored fruiting bodies with a distinct subspore apophysis (Figs. 12–17). *Cavostelium bisporum* Olive & Stoian. (1966c), which is also widespread but less common on similar substrates, consistently has 2-spored sporocarps with smaller apophysis (Fig. 19). There is a distinct hilar area on the side of each spore where it comes into contact with its partner (Fig. 24). Recently, the spore walls of *C. apophysatum,* like those of *C. bisporum,* have been found to have fine punctate markings. A sporocarp becomes 2-

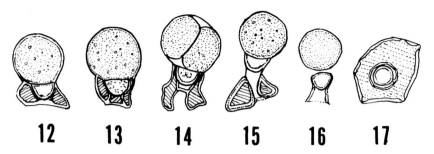

| 12 | 13 | 14 | 15 | 16 | 17 |

Figs. 12–17. Sporophores of *Cavostelium apophysatum*, showing gas-inflated sheath, apophysis, spores, and view of stalk and stalk base from above (Fig. 17), ×1400. From Olive (1964a).

spored by mitotic cell division of the uninucleate sporogen at the conclusion of stalk formation. Bacteria are ingested as food (Figs. 20–22).

Ceratiomyxella Olive & Stoian. (1971a) is a monotypic genus characterized by a more varied life cycle than is found in most protostelids. *Ceratiomyxella tahitiensis* Olive & Stoian. was first found on dead, moldy, still attached fruit pedicels of a grapefruit tree in Tahiti, and a variety of it (var. *neotropicalis*) was later found on leguminous pods in Brazil and on dead grass in Colombia. Its life cycle is diagrammed in Fig. 25. The deciduous, plurinucleate spores are borne singly on slender, hollow stalks (Figs. 25M and 26). On germination each spore produces a plurinucleate protoplast that soon becomes lightly encysted to form the zoocyst, or two or three zoocysts if the protoplast is large (Fig. 25, A and B). Before encystment, protoplasts of different spores may also fuse with each other. All nuclei except one in the zoocyst degenerate (Fig. 25B). The remaining nucleus enlarges, usually undergoes three successive divisions, and eight uninucleate flagellate cells are delimited (Figs. 25, C–F, and 27). In a number of ways these developments resemble those that occur during spore maturation and germination in *Ceratiomyxa*, except that the first two divisions, known to be meiotic, occur in the spores of *Ceratiomyxa*, and the flagellate cells are reported to function as gametes. Although meiosis and gametic cell fusions have not been demonstrated in *Ceratiomyxella tahitiensis*, the limited observations on this species do not preclude their occurrence. If meiosis does occur here, the most logical place to suspect it is in the zoocysts.

Single-cell cultures of *C. tahitiensis*, including those from single flagellate cells formed in the zoocysts, give rise to mature plasmodia and sporocarps (Fig. 25, F–M). The plasmodia become reticulate (Figs. 25K and 29) and their nuclear division tends to be synchronous, at least in certain areas (Fig. 30). There is little information on the details of mitosis in the plasmodium except for an isolated observation that

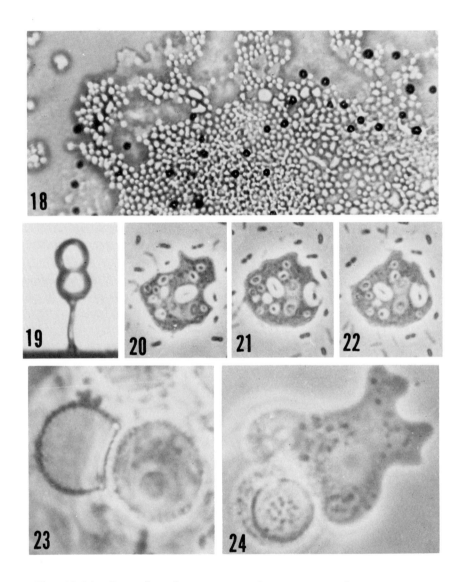

Figs. 18–24. *Cavostelium bisporum.* Fig. 18. Sporocarps and cysts, ×200. Fig. 19. Sporocarp, ×1250. Figs. 20–22. Stages in ingestion of a bacterial cell by a binucleate amoeboid cell. Phase contrast ×1850. Fig. 23. Spore pair showing roughened walls (one spore has germinated), ×3300. Fig. 24. Germinating spore, showing hilar region and punctate wall markings, ×3300. Figs. 19–22 from Olive and Stoianovitch (1966c).

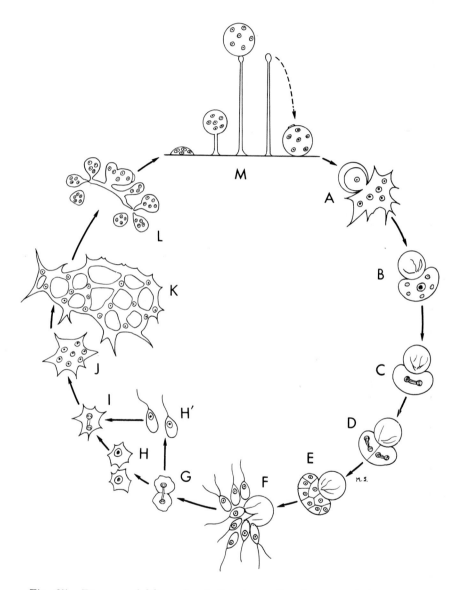

Fig. 25. Diagram of life cycle of *Ceratiomyxella tahitiensis:* (A) germinating spore; (B–F) stages in zoocyst development; (G, H, H') proliferation of amoeboid and flagellate cells; (I, J) early stages in plasmodial development; (K) reticulate plasmodium; (L) prespore cell formation; (M) stages in sporogenesis. From Olive (1970).

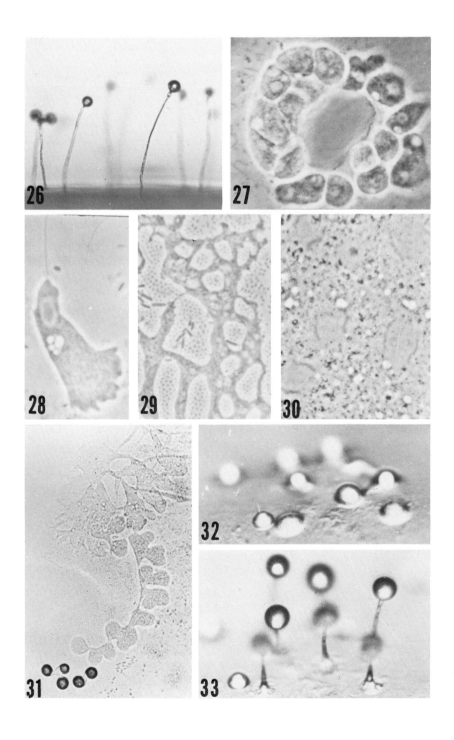

a distinct spindle is present, with no evidence of a persistent nuclear membrane around it (Furtado and Olive, 1971a). Prior to sporulation, the plasmodium becomes lobed and then segments into plurinucleate prespore cells that quickly become hemispherical and then opaque as the covering slime sheath is secreted (Figs. 25L and 31). Sporulation then proceeds as previously described (Figs. 32 and 33). As in *Nematostelium*, there is a distinct apophysis at the tip of the stalk to which the spore is attached by a hilar area (Figs. 25M and 26). This arrangement is apparently related to the deciduous character of the spore. The stalks give a positive reaction for cellulose with the iodine–potassium iodide–sulfuric acid test.

Various-sized, mostly plurinucleate segments of the plasmodia may secrete thickened walls and become resting cysts. They may readily be distinguished from zoocysts by their thicker walls, greater variability in size and shape, and failure to germinate readily.

Planoprotostelium Olive & Stoian. (1971b) is also a monotypic genus, known only by a single isolate from dead papyrus inflorescences collected in Brazil. *Planoprotostelium aurantium* Olive & Stoian. closely resembles *Protostelium mycophaga*, both in the orange pigmentation and similar single-spored fruiting bodies produced in great profusion when cultured on the pink yeast *Rhodotorula*. However, the spores of *Planoprotostelium* are mostly nondeciduous and, perhaps more important, the uninucleate amoeboid cells of the trophic stage become flagellate in water. Cells with supernumerary flagella are very common, with up to seven flagella occurring on a single cell. Only one long flagellum has been found associated with the nucleus; a short, reflexed flagellum has not been detected.

The question of the importance of flagellation as a major taxonomic character in classification has already been discussed. This problem cannot be solved until we are better able to assess the relative values of other taxonomic criteria as well. More information on the structure and life histories of the protostelids is particularly required.

Protosporangium Olive & Stoian. (1972) has minute, mostly 2- to 4-spored nondeciduous microsporangia borne on unusually slender, flexu-

---

Figs. 26–33. *Ceratiomyxella tahitiensis.* Fig. 26. Sporocarps on agar medium, ×210. Fig. 27. Zoocyst containing eight cells; eight flagellate cells have been released from second zoocyst; empty spore wall in center. Phase contrast ×1250. Fig. 28. Single flagellate cell. Phase contrast ×1360. Fig. 29. Reticulate plasmodium. Phase contrast ×515. Fig. 30. Synchronous nuclear division (anaphase) in plasmodium. Phase contrast ×1250. Fig. 31. Prespore cells segmenting from plasmodium, ×210. Figs. 32 and 33. Stages in sporogenesis, ×435. From Olive and Stoianovitch (1971a).

ous stalks. The spore walls are smooth. The trophic stage consists of uninucleate to plurinucleate protoplasts but never reticulate plasmodia. The plurinucleate plasmodia often assume a wormlike shape (Fig. $34G_2$), the significance of which is unknown. Plurinucleate protoplasts segment into uninucleate prespore cells that give rise to the sporo-

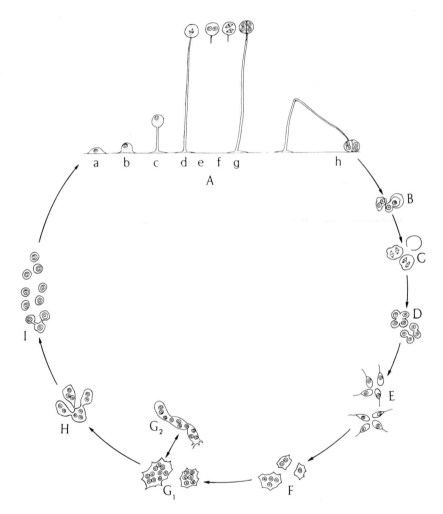

Fig. 34. Life cycle of *Protosporangium bisporum*: (A) stages in sporogenesis; (B) spore germination; (C) mitosis in spore protoplasts; (D, E) segmentation of the two 4-nucleate protoplasts to form eight flagellate cells; (F, G) development of plurinucleate protoplasts; vermiform phase shown in G_2; (H, I) segmentation of plurinucleate protoplast to form uninucleate prespore cells. From Olive and Stoianovitch (1972).

carps (Fig. 34A and G–I), although in one species single amoeboid cells may also convert directly into prespore cells. At the end of stalk elongation, two successive mitoses are generally believed to occur in the uninucleate sporangial protoplast prior to spore delimitation. The spores germinate to produce flagellate cells. The total number of flagellate cells that arise from the germination of all spores of a single sporangium (whether two or four) is usually eight. Attempts at single-spore cultures have failed. The species of this genus must be further investigated to determine whether a sexual process is present and whether the two nuclear divisions in the young sporangia are meiotic.

Protosporangium bisporum Olive & Stoian. (Fig. 35), found on the bark of trees in North Carolina and Uganda, has sporangia with usually two binucleate spores but sometimes only one. Its life cycle is illustrated in Fig. 34. Each spore of a binucleate sporangium germinates to produce a binucleate protoplast in which the two nuclei divide conjugately (Figs. 34, B and C, and 38). The resultant 4-nucleate protoplast then becomes four-lobed and divides into four flagellate cells (Figs. 34, D and E, and 39). When there is only a single spore, it is 4-nucleate, and when its protoplast emerges from the spore a synchronous nuclear division makes it 8-nucleate, after which it becomes lobed and segments into eight flagellate cells (Figs. 39 and 40).

Protosporangium fragile Olive & Stoian., found on rotting pine wood in North Carolina, has sporangia with four uninucleate spores. At germination a single mitosis occurs in the emergent protoplast which then divides into two flagellate cells, each of which has a single long flagellum and an inconspicuous, short reflexed one, or a pair of long flagella (Fig. 41). This species was revived on its original substrate after storage for 16 months at 17°C.

Protosporangium articulatum Olive & Stoian. (Figs. 36 and 37) has been found on tree bark collected in North Carolina, Michigan, France, and England. The sporangia commonly have two to four spores, but sometimes in culture as many as eight spores may be found. The spores are uninucleate and germinate as do those of *P. fragile*. The variability in spore number is probably related to the fact that some prespore cells may contain two or even three nuclei and to the probable occurrence of some nuclear degeneration in the sporangia.

The relationships of *Protosporangium* to other protostelids is uncertain. The common development of eight flagellate cells from the spores of a single sporangium shows some resemblance to the development of the same number of flagellate cells from the spore protoplast of *Ceratiomyxa*. Also, the wormlike protoplasts found in *Protosporangium* (Fig. 34G$_2$) are known elsewhere only in *Ceratiomyxa*, the emergent spore

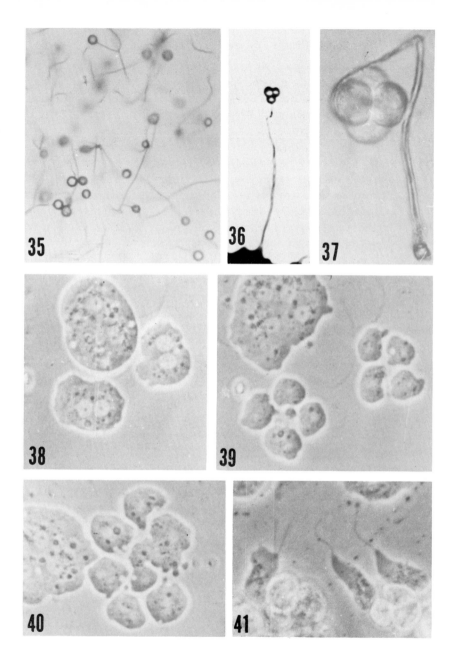

Figs. 35–41. *Protosporangium*. Fig. 35. Sporocarps of *P. bisporum* on agar surface, ×275. Fig. 36. Four-spored sporocarp of *P. articulatum*, ×275. Fig. 37. Sporocarp enlarged, showing tubular stalk, ×1325. Figs. 38–40. Stages in development of flagellate cells from binucleate and 4-nucleate spore protoplasts in *P. bisporum*. Phase contrast ×1900. Fig. 41. Flagellate cells of *P. fragile*. Phase contrast ×1250. Figs. 35–37 and 39 from Olive and Stoianovitch (1972).

protoplasts of which often assume a similar shape (thread phase) before they divide into flagellate cells. Comparisons may also be made with the myxomycete genus *Echinostelium*, which contains the most diminutive slime molds, the smallest of which is *E. lunatum* Olive & Stoian. (1971c) with 4- to 8-spored sporangia. However, the *Echinostelium* sporangium develops from a single protoplasmodium rather than from prespore cells that typically develop by segmentation of plurinucleate protoplasts. In addition, as has been noted, the spore wall markings of *E. lunatum* indicate its origin from the protostelid genus *Cavostelium*. Therefore, it is likely that *Protosporangium*, in spite of certain resemblances to myxomycetes, is not in the main line of evolution of myxomycetes from protostelids.

B. Ceratiomyxidae Schroeter (1897)

This family has generally been classified with the myxomycetes, the tradition having been established long before the discovery of the simpler protostelids. The single genus *Ceratiomyxa* is characterized by microscopic single-spored, slender-stalked sporocarps borne on raised plasmodial columns that solidify on drying. The genus has always seemed an anachronism in company with the endosporous Myxogastria. Because of the striking similarities between the plasmodia and sporocarps of *Ceratiomyxa* and those of some of the simpler protostelids, the genus along with its family were recently transferred to the Protosteliida (Olive, 1970). Martin and Alexopoulos recognize three species.

The best known and most frequently studied species is *C. fruticulosa* (Müll.) Macbride (Fig. 42), which is widely distributed from the arctic to the tropics on decaying wood. Prior to fruiting, the nearly hyaline plasmodium embedded in a copious matrix of mucus flows out onto the surface of a dead log and turns milky white. Microscopic examination shows that the plasmodium is reticulate (Fig. 43). Typically, the plasmodium and its mucous matrix become extended into upright, usually branched, white columns 1–10 mm tall. Sometimes raised poroid structures are formed instead and color variations are also known. The plasmodial network, in which the shaping forces must reside, assumes a peripheral position in the columns, followed by a synchronous nuclear division (Gilbert, 1935), and segments into uninucleate prespore cells ("protospores") as in certain other protostelids (Figs. 44 and 45). The prespore cells then give rise to slender-stalked sporocarps, each bearing a single round to cylindrical spore, in typical protostelid fashion (Figs. 44–46). Occasionally, collections are found that lack spore columns and the

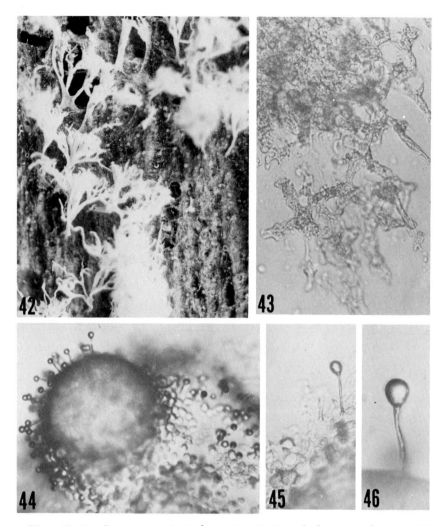

Figs. 42–46. *Ceratiomyxa fruticulosa.* Fig. 42. Branched spore columns, ×15. Fig. 43. Portion of plasmodium showing its reticulate nature, ×300. Fig. 44. Portion of spore column showing individual sporocarps and prespore cells, ×200. Fig. 45. Sporocarp and prespore cells, ×300. Fig. 46. Single sporocarp, ×700.

sporocarps arise from a nearly flat or, at most, bumpy surface. This more closely resembles the condition of fruiting in other protostelids.

A sheath surrounds each tiny sporocarp and, as E. W. Olive (1907) noted, the slender stalk tube is hollow. Both Famintzin and Woronin (1873), who first described and beautifully illustrated its development

(Fig. 47), and E. W. Olive observed that during sporogenesis a portion of the sporogen, probably homologous with the steliogen of other protostelids, extends into the top of the stalk. As the sporocarps mature, the mucilaginous matrix of the spore columns dries and the latter become firm supporting structures for the many sporocarps that cover their surfaces. The spores are deciduous and readily fall off their stalks.

The major impediment to a better understanding of the life cycle of *C. fruticulosa* is its resistance to culture in the laboratory. Gilbert (1935) reported observing reversible flow of protoplasm in a plasmodium collected on the natural substrate, but this could not be confirmed by Olive (1970), who concluded that the alternate ebb and flow characteristic of reticulate myxomycete plasmodia does not occur in *Ceratiomyxa*. Possibly, Gilbert was observing an associated myxomycete plasmodium.

Although the spore of this species is at first uninucleate, it eventually becomes 4-nucleate from two successive nuclear divisions that may take place before or after the spore is shed. Several investigators (e.g., E. W. Olive, 1907; Sansome and Dixon, 1965) have considered these divisions to be meiotic, with the haploid chromosome number being eight. The discovery of synaptonemal complexes in the segmenting prespore cells (Furtado and Olive, 1971b) supports this concept and indicates that meiotic prophase in *Ceratiomyxa* begins at an earlier stage than in myxomycetes, in which meiosis appears to be confined entirely to the spore after its delimitation in the sporangium.

As the 4-nucleate protoplast emerges from the spore during germination, it often enters a wormlike or "thread" phase (Gilbert, 1935). Another nuclear division ensues and the resultant 8-nucleate protoplast

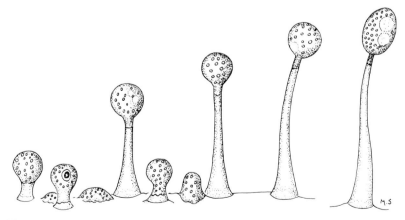

Fig. 47.   Sporogenesis in *Ceratiomyxa fruticulosa*. Redrawn from Famintzin and Woronin (1873).

becomes eight-lobed and divides into eight flagellate cells exactly as do protoplasts of single-spored fruiting bodies of *Protosporangium bisporum*. The flagellate cell has one long flagellum, with or without a short reflexed flagellum apparent at its base.

Gilbert (1935), McManus (1958), and Sansome and Sansome (1961) have reported plasmogamy between flagellate cells, but these accounts have often failed to distinguish clearly between plasmogamy and delayed separation of flagellate cells from the original 8-nucleate protoplast. However, Gilbert reports having observed nuclear fusion following plasmogamy. McManus was unable to confirm this. The Sansomes have obtained small incipient plasmodia in culture, but attempts to propagate the organism in the laboratory beyond this point have failed. Thus far, *C. fruticulosa* remains the only protostelid in which sexual reproduction has been found, although it may yet be found in other known members of the subclass.

Two other species, tropical in distribution, are recognized by Martin and Alexopoulos (1969). They differ from *C. fruticulosa* in having smaller spore-bearing extensions of the plasmodium and somewhat smaller spores. *Ceratiomyxa sphaerosperma* Boedijn has subglobose, slightly roughened spores. *Ceratiomyxa morchella* Welden has somewhat morchelloid spore-bearing structures and smooth, oval to elliptical spores. Sometimes the spore-bearing surface remains effuse and only folded.

C. Protosteliidae Olive & Stoian. (1966b)

This family includes all protostelids that lack a flagellate stage. They appear to have evolved more than once from flagellate protostelids by loss of flagella. As noted earlier, the presence or absence of flagella is of dubious value in determining closeness of relationship in these organisms. If this concession is kept in mind, the use of this characteristic for purposes of taxonomic convenience can perhaps be tolerated until we know more about the group and can decide which characteristics are more salient. The family may be segregated into those with deciduous spores and those without.

The deciduous-spored protostelids—*Schizoplasmodium, Nematostelium,* and *Protostelium*—have a subspore apophysis that ranges from distinct to much reduced. *Schizoplasmodium cavostelioides* Olive & Stoian. (1966a), the only known species in the genus, has sporocarps that are very similar in appearance to those of *Cavostelium apophysatum*. The stalk is very short and the single spore is subtended by a distinct apophysis. However, the trophic stage is unlike that of *Cavostelium* in being a reticulate plasmodium that segments just before sporulation into plurinucleate prespore cells. The spores are also plurinucleate. The most remarkable feature of this species is its ballistospores, the only ones known in the Mycetozoa. As soon as the sporocarp matures, a gas bubble appears to one side of the spore between spore wall and sheath. The latter is stretched until the bubble reaches a diameter ranging from about half that of the spore to somewhat larger than the spore, after which it bursts and discharges the spore to a distance horizontally of up to 200 μ, leaving the vacated stalk and apophysis behind (Figs. 48–50). The same mechanism has been claimed to motivate spore discharge in basidiomycetes (Olive, 1964b). At germination the plurinucleate protoplast emerges from the spore case and develops directly into a reticulate plasmodium. The species is common in various parts of the world, including the southeastern United States, on moribund, attached flowers, fruits, and grass inflorescenses. It grows and sporulates well in culture on an unidentified yeast.

Schizoplasmodiopsis has reticulate plasmodia that resemble those of

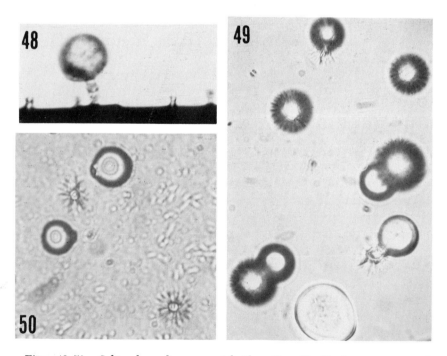

Figs. 48–50. *Schizoplasmodium cavostelioides.* Fig. 48. Single sporocarp and vacated stalks on agar surface, ×560. Fig. 49. Sporocarps with gas bubbles, ×650. Fig. 50. Discharged spores (note hilum) and vacated stalks, ×650. From Olive and Stoianovitch (1966a).

Schizoplasmodium and that segment to form prespore cells, which give rise to small single-spored sporocarps. The spores are nonapophysate and nondeciduous. The single known species is *S. pseudoendospora* Olive, Martin & Stoian. (Olive, 1967), illustrated in Figs. 51–53. The specific name refers to the presence of sporelike but enucleate bodies within the cysts. Both prespore cells and spores are uninucleate and the stalks of the sporocarps are relatively short. The species is not a common one; it has been isolated only twice from dead flowers and capsules collected in Florida.

Nematostelium Olive & Stoian. (Olive, 1970) is recognized by its long-stalked sporocarps with distinctive apophysate swelling at the stalk apex (Fig. 55). The knoblike apical enlargement becomes more apparent after the plurinucleate spore falls off. A circular hilum is found on that part of the spore which was attached to the apophysis. The trophic stage is primarily a reticulate plasmodium (Fig. 56). On agar media the plasmodium often produces extensive, branched, filose pseudopodia (rhizopodia) that extend down into the medium and at times anastomose with each other (reticulopodia). Encysted protoplasts range from small rounded cells to larger irregular bodies (Fig. 57). Two species of rather widespread distribution are known. *Nematostelium ovatum* (O. & S.) Olive & Stoian., with ovate spores (Figs. 11 and 54–57), occurs in soil and humus, whereas *N. gracile* (O. & S.) Olive & Stoian., with round spores, is found primarily on dead attached plant parts. *Ceratiomyxella tahitiensis* has sporocarps that closely resemble those of *N. gracile* but

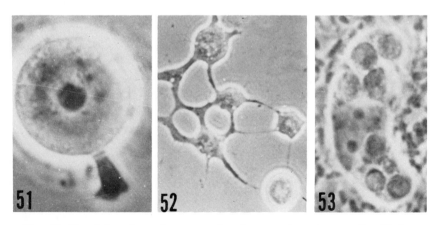

Figs. 51–53. *Schizoplasmodiopsis pseudoendospora*, phase contrast. Fig. 51. Sporocarp with uninucleate spore, ×2900. Fig. 52. Reticulate plasmodium delimiting prespore cells, ×1200. Fig. 53. Binucleate cyst with sporoid bodies, ×2900. From Olive (1967).

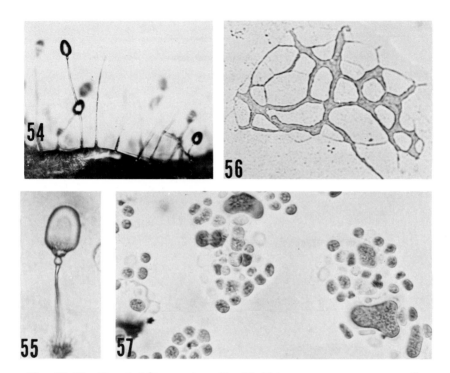

Figs. 54–57. *Nematostelium ovatum*. Fig. 54. Mature sporocarps on agar surface, ×200. Fig. 55. Sporocarp enlarged; note "ball-and-socket" articulation of spore with apophysis at stalk tip, ×650. Fig. 56. Reticulate plasmodium, ×650. Fig. 57. Encysted protoplasts, ×650. From Olive and Stoianovitch (1966b).

the former species may be distinguished by its zoocysts and flagellate cells. It is likely that *Nematostelium* evolved from *Ceratiomyxella*.

Protostelium, thus far the largest genus of protostelids, has been monographed by Olive and Stoianovitch (1969). There are five known species, most of which are found on dead attached plant structures. The trophic stage consists primarily of uninucleate amoebae, with or without plurinucleate protoplasts intermixed. The individual uninucleate amoebae function as prespore cells to produce slender-stalked, single-spored fruiting bodies. The subspore apophysis is much reduced and may accompany the spore when it falls off the stalk.

Probably the most common species (and possibly the most common protostelid) is the orange-pigmented *P. mycophaga* Olive & Stoian. It grows and sporulates well on agar media when supplied with any of several yeasts or conidia of certain fungi. It also grows and fruits on killed yeast cells. Three varieties have been described (Olive and Stoia-

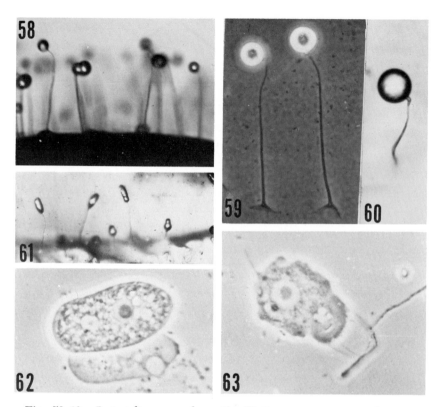

Figs. 58–60. *Protostelium mycophaga.* Fig. 58. Sporocarps on agar surface, ×300. Fig. 59. Sporocarps, showing reduced apophysis and cupulate disc at stalk base. Phase contrast ×600. Fig. 60. Sporocarp showing attachment of stalk and subspore apophysis, ×720. Figs. 58 and 59 from Raper (1960); Fig. 60 from Olive and Stoianovitch (1969).

Figs. 61–63. *Protostelium arachisporum.* Fig. 61. Sporocarps on agar surface, ×200. Fig. 62. Two spores (note hilum at base of squashed spore). Phase contrast ×1160. Fig. 63. Germinating spore. Phase contrast ×1160.

novitch, 1969). The type variety has sporocarps up to about 75 μ tall and spores that are 9–14 μ in diameter (Figs. 58–60). The reduced apophysis may usually be found attached to the fallen spore. Heavily sporulating cultures, when inverted, produce orange spore prints. The trophic stage is also orange pigmented, a characteristic that can be dramatically illustrated by growing it on a white yeast. Light is not required for fruiting and no evidence of phototaxis has been observed. Probably the closest relative of *P. mycophaga* is *Planoprotostelium aurantium*, from which it differs primarily in the absence of flagellate cells.

Protostelium irregularis Olive & Stoian. produces some plurinucleate protoplasts in addition to uninucleate amoebae. The amoebae are typically fan-shaped when migrating. The round spores readily fall off the slender stalks, which may then be distinguished from those of other protostelids by the slight spindle-shaped apical swelling. Some of the sporocarps produce two spores. The microcysts are often irregular in shape. *Protostelium arachisporum* Olive & Stoian. has ovate to peanut-shaped spores (Figs. 61–63), whereas *P. zonatum* Olive & Stoian. (Figs. 64 and 65) and *P. pyriformis* Olive & Stoian. have bell-shaped to pyriform spores, those of the latter species being distinctly smaller. In both of the latter, the stalk is formed in a manner unlike that in any other protostelid. The developing apical end of the stalk is enclosed by a cylindrical steliogen within the culminating protoplast (Fig. 65). In *P. zonatum* minute vesicles are laid down in a single series within the stalk, giving it a zonate appearance that makes the species easily recognizable even under relatively low magnification. An ultrastructural

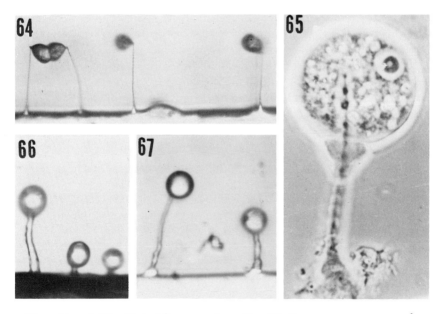

Figs. 64 and 65. *Protostelium zonatum.* Fig. 64. Sporocarps on agar surface, ×200. Fig. 65. Developing sporocarp showing stalk within the steliogen, the latter extending below the main body of the protoplast. Phase contrast ×1400. From Olive and Stoianovitch (1969).

Figs. 66 and 67. *Protosteliopsis fimicola.* Sporocarps on agar surface, ×840. From Olive and Stoianovitch (1966c).

study of sporocarp morphogenesis in this species should prove especially interesting. All species of *Protostelium* are widely distributed.

Protosteliopsis is known by the single species *P. fimicola* (Olive) Olive & Stoian. (1966d), which has a trophic stage consisting chiefly of uninucleate amoebae. The round uninucleate spores, which are borne on stalks of moderate length, are nonapophysate and nondeciduous (Figs. 66 and 67). The amoebae have pseudopodia that are less pointed than in most protostelids. The stalks tend to deliquesce readily in aqueous mounts. This is a common species throughout the world on dung of various types, in soil, and on dead plant structures.

VI. ULTRASTRUCTURE

Fine structure studies of several protostelids, including members of the flagellate genera *Cavostelium* (Figs. 68–70) and *Ceratiomyxella* (Figs. 73 and 75) (Furtado and Olive, 1970a, 1971a, 1972; Furtado *et al.*, 1971) and the nonflagellate genera *Protostelium* and *Protosteliopsis* (Figs. 71 and 72) (Hung and Olive, 1972b, 1973a,b), reveal many similarities with most other mycetozoans. Pseudopodia and other hyaloplasmic areas of amoeboid cells have a fibrous substructure and lack major cell organelles (Figs. 71 and 72). The mitochondria typically have tubular, or sometimes vesicular, cristae. Limited amounts of rough endoplasmic reticulum are found in the cytoplasm and ribosome-free Golgi stacks occur in amoebae and flagellate cells. Food cells are taken in by ingestion (Fig. 72). Conspicuous food vacuoles contain cells of the food organism and sometimes cannibalized cells of the protostelid in various stages of digestion. In later stages of digestion the food residues often accumulate as a series of concentric membranous elements (myelin figures), which may be found outside the cell following discharge of the food vacuoles. Autophagic vacuoles, which contain ingested cytoplasm and cell organelles, are also found at certain stages. Pinocytotic vacuoles, lacking solid contents, and contractile vacuoles are present in trophic stages. After expelling its clear contents, the contractile vacuole reorganizes by the fusion of many smaller vacuoles in the cytoplasm. The nucleus is typical in structure and has a single central nucleolus. Observations under the electron microscope on nuclear division in protostelids are few and it is possible to state with certainty only that chromosomes are present on a spindle of microtubules.

Studies on two flagellate protostelids, *Cavostelium bisporum* (Furtado and Olive, 1970a) and *Ceratiomyxella tahitiensis* (Furtado and Olive, 1971a), have demonstrated that the flagellar apparatus is basically like

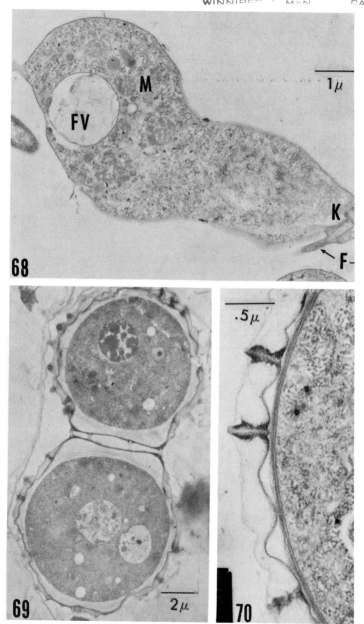

Figs. 68–70. Electron micrographs of *Cavostelium bisporum*. Fig. 68. Thin section of flagellate cell showing food vacuole (FV), mitochondria (M), a pair of kinetosomes (K), and a pair of flagella (F). The reflexed flagellum is the shorter one. Fig. 69. Spore pair, showing channelized structures passing through spore wall, connections between spore hila, and reencysted spore protoplasts. Fig. 70. Enlargement of channelized wall structures. Fig. 68 from Furtado and Olive (1970a); Figs. 69 and 70 from Furtado *et al.* (1971).

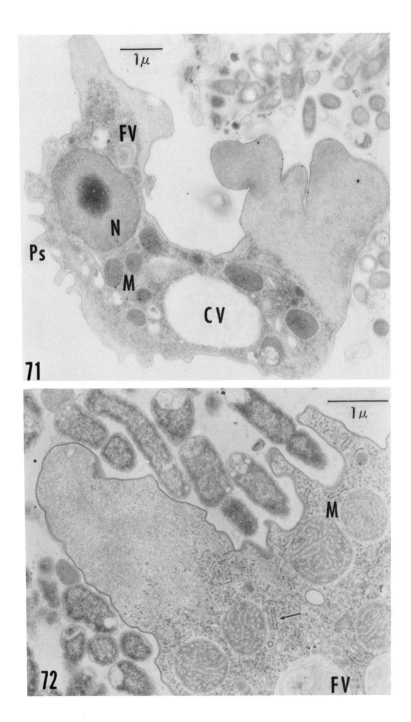

that of myxomycetes (Fig. 68). A pair of kinetosomes arranged in a V shape or at right angles give rise to arrays of microtubules that pass back into the cytoplasm. In *C. bisporum* these microtubules occur in a conelike arrangement delimiting an inner cytoplasmic area between kinetosomes and nucleus that contains Golgi and abundant endoplasmic reticulum. A single long flagellum arises from one kinetosome and a short reflexed one from the other (Fig. 68). Under the phase-contrast microscope the short flagellum is seen closely applied against the plasmalemma near the base of the long flagellum. It could have little if any influence on cell propulsion but might assist in stabilizing the flagellar apparatus. Flagella, kinetosomes, and centrioles are typical in structure. Some evidence of centriole proliferation in clusters has been found in *Ceratiomyxella tahitiensis*, and, as the centrioles are thought to function alternately as kinetosomes, this may help explain the occurrence of supernumerary flagella in this and other flagellate protostelids.

A most interesting development, so far not reported in any other mycetozoan, occurs in the amoeboflagellate cells of *C. tahitiensis* (Furtado and Olive, 1971a, 1972). Within flattened vesicles (Fig. 74), probably derived from the Golgi apparatus, oval scales (Fig. 75) develop and are subsequently extruded at the periphery of the cell. In sectional view the scales appear saucer-shaped with recurved edges, and they come to line the exterior of the cell in several layers with their concave sides facing outward (Fig. 73). Their function is unknown. Similar structures have previously been reported only in certain algal protists (see Furtado and Olive, 1971a) and in members of the Labyrinthulina (see Chapter 7). This discovery strongly suggests that the protostelids have evolved from algal protists.

Encystment of the amoeboflagellate cells has been studied in *Cavostelium bisporum* (Furtado and Olive, 1970b). It is preceded by the disappearance of flagella and most of the food vacuoles. Pseudopodia cease to develop and a slime sheath is secreted around the cell. Vesicles appear in the cytoplasm, move to the periphery of the cell, and release their contents after fusing with the plasmalemma. It is likely that they contribute to formation of the cyst wall, which accumulates in thin successive layers. The kinetosomes withdraw from their peripheral position and

---

Figs. 71 and 72. Electron micrographs of *Protosteliopsis fimicola*. Fig. 71. Uninucleate amoeboid cell. Note pseudopodia (Ps), fibrillar hyaloplasmic areas of the cell, nucleus (N), mitochondria (M), food vacuole (FV), and contractile vacuole (CV). Fig. 72. Portion of amoeboid cell, showing ingestion of bacterial cells. Note row of mitochondria (M), food vacuole (FV), short segment of rough ER (arrow), and fibrillar hyaloplasmic area at left. From Hung and Olive (1972b).

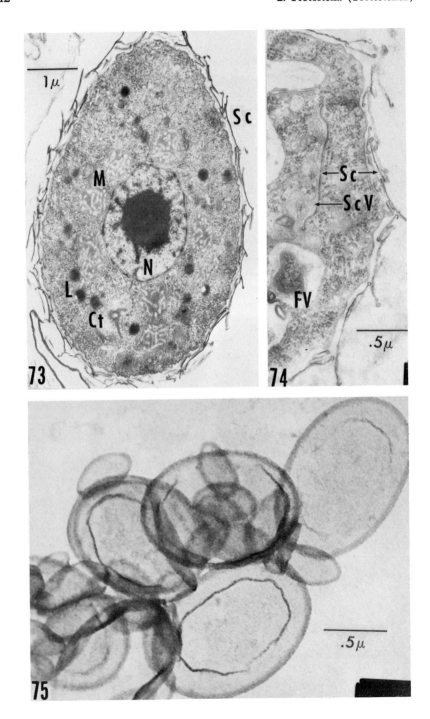

gradually lose their associated microtubules. The process of encystment closely resembles that in myxomycetes.

Only the fruiting bodies of *Cavostelium bisporum* have been studied in any detail under the electron microscope (Furtado *et al.*, 1971). The presence of a distinct supporting disc at the base of the stalk has been confirmed. The minute punctate markings that are seen on the spore walls under the phase-contrast microscope have proved to be the tips of channelized structures that extend through the breadth of the wall and emerge at the surface (Figs. 69 and 70). The hilar areas between spore pairs are connected by peculiar wall structures that may be of similar origin. The channelized wall processes are known elsewhere in mycetozoans only in the myxomycete genus *Echinostelium,* two species of which have been shown to have them (Hung and Olive, 1972a). This is of particular interest phylogenetically because these are also the simplest, and probably the most primitive, myxomycetes.

The electron microscope has also confirmed the presence of a sheath as the limiting outer element of the sporocarps. As in microcyst development, the kinetosomes withdraw into the spore cytoplasm, where their associated microtubules disappear. No nuclear division has been detected in the spores of *Cavostelium* or in any other protostelid prior to germination, except in some collections of *Ceratiomyxa fruticulosa.*

The fine structure study by Scheetz (1972) on sporulation in *C. fruticulosa* confirms Gilbert's report of nuclear division (probably synchronous) in the plasmodium of the spore column prior to formation of the prespore cells. The division appears to be intranuclear. Furtado and Olive (1971b) have found synaptonemal complexes in prespore cells at the time of their delimitation, which tends to confirm several earlier reports of sexuality in this species. Scheetz has found the two successive nuclear divisions in the spore to be intranuclear. Four centrosomes, four nuclei, and a Golgi body associated with each nucleus were found in the mature spore. As noted above, nuclear divisions in the spores are probably meiotic and in some collections they occur after the spores are shed. Scheetz does not think that his findings indicate a closer relationship of *Ceratiomyxa* to protostelids than to myxomycetes.

Figs. 73–75.   Electron micrographs of *Ceratiomyxella tahitiensis.* Fig. 73. Amoeboflagellate cell surrounded by scales, one still in elongate scale vesicle. Fig. 75. Whole mount of scales. Sc, scales; ScV, scale vesicle; N, nucleus; M, mitochondria; Ct, centriole; L, lipid bodies. From Furtado and Olive (1971a).

CHAPTER 3

Dictyostelia
(Dictyostelid Cellular Slime Molds)

I. INTRODUCTION

Brefeld (1869) discovered the first cellular slime mold and named it *Dictyostelium mucoroides,* the generic name referring to the network of empty cells in the stalk of the fruiting body and the specific name to the resemblance of the fruiting bodies to the sporophores of certain mucoraceous fungi. Although Brefeld thought that a true plasmodium was produced, van Tieghem (1880) demonstrated that the aggregate of cells formed prior to fruiting retained its cellularity. Therefore, he called it a "pseudoplasmodium." Van Tieghem also created the order Acrasieae to accommodate both dictyostelid and acrasid cellular slime molds. De Bary (1887) recognized the Acrasieae and Myxomycetes as orders of coordinate rank in his newly formed group, the Mycetozoa. In his monograph of all the cellular slime molds known at the time, E. W. Olive (1902) retained the single order Acrasieae, in which he recognized three families: Sappiniaceae, Guttulinaceae, and Dictyosteliaceae. Reasons for excluding the Sappiniaceae and for widely separating the other two groups were discussed in Chapter 1. As recently recommended (Olive, 1970), the dictyostelid cellular slime molds are here treated as a separate subclass of the Eumycetozoa, the Dictyostelia, containing the single order Dictyosteliida and two families, Acytosteliidae and Dictyosteliidae.

Studies on the cellular slime molds over a period of nearly 40 years by Kenneth Raper and his co-workers have contributed immensely to our understanding of these organisms and have led to their increasing use in various aspects of developmental research. Because the dictyostelids are relatively simple and malleable subjects for such investigations

45

and because their developmental stages occur in well-defined series, they have become increasingly popular laboratory subjects for research on gross morphology, ultrastructure of morphogenesis, and biochemical analysis, and the literature on them has rapidly proliferated. Bonner (1967, 1971) has published useful, comprehensive surveys on the biology of cellular slime molds, with emphasis on the better known dictyostelids, and Gregg (1964, 1966) and Newell (1971) have reviewed the biochemical investigations of morphogenesis.

II. OCCURRENCE, ISOLATION, AND MAINTENANCE

The majority of dictyostelid cellular slime molds have been isolated from forest soil and humus and from the dung of various animals in both temperate and tropical zones. The greatest diversity of species occurs in tropical and subtropical zones, where at least six of them are endemic (Cavender, 1973). Raper (1951) has described methods for their isolation and maintenance. This and other studies (e.g., Cavender and Raper, 1965) have shown that variety and frequency vary with the type of forest, relatively dry hickory–oak associations being most productive in North America. In our search for protostelids we have also found dictyostelids on such additional substrates as rotting wood, decaying fungi, and moribund attached plant structures (pods, capsules, fleshy fruits, etc.). The sorocarps are inconspicuous in nature and are generally not observed until isolated in the laboratory. However, early on a dewy morning we have found sorocarps of *Dictyostelium mucoroides* arising from a group of rabbit pellets on the ground in such abundance that they appeared as a whitish moldlike growth on the surface. On another occasion, it was possible to detect with a hand lens the compound sorocarps of *Polysphondylium pallidum* on decaying mushrooms growing on wood.

An effective technique for isolating these organisms in the laboratory has been described (Cavender and Raper, 1965). Suspensions of the substrate mixed with suspensions of a suitable food bacterium are plated out on hay infusion agar. A wide variety of bacteria, including some pathogenic ones can be utilized by the dictyostelids, but none has proved superior to *Escherichia coli* or *Aerobacter aerogenes*. The above technique permits the recording of numbers of clones on the agar plates and thus yields information on relative abundance of dictyostelids in the substrate. The weakly nutrient agar and the bacterial growth initiated on its surface greatly reduce the growth of unwanted fungi. When dictyo-stelid sorocarps appear, it is a simple matter to pick up on the tip of a sterile needle a mucilaginous spore mass and transfer it to a fresh plate of medium, where the desired bacterium is added.

Dictyostelids are easily grown in the laboratory with a bacterium on a variety of culture media, such as hay infusion, lactose–yeast extract, and peptone–dextrose agars (see Appendix). A temperature of 21°–25°C and a pH of about 6 are usually considered optimal. Most species fruit readily in a few days on these media. They can be maintained in the laboratory with little effort if transferred every month or so, or their spores may be lyophilized and kept in viable condition over a period of years.

Large numbers of amoebae for experimental purposes may be obtained in a liquid medium containing yeast extract, peptone, dextrose, and a few minerals if flasks of the medium are inoculated with the food bacterium and placed on a shaker for 48 hours before spores of the dictyostelid are added (Sussman, 1961a). Several species can grow on dead bacteria. Strain WS-320 of *Polysphondylium pallidum* is the best performing one under these conditions (Hohl and Raper, 1963). This same strain has been successfully grown in cell-free liquid medium (Hohl and Raper, 1963) and finally on a defined liquid medium containing sucrose, glycerol, acetate, lactate, citrate, glutamate, and several amino acids (Goldstone *et al.*, 1966). These axenic culture techniques enlarge the scope of investigations possible with such a dictyostelid because they eliminate the food organism and facilitate biochemical and genetic studies. Only recently have suitable strains of *D. discoideum* been ob-tained that can be grown axenically (Watts and Ashworth, 1970; Wil-liams *et al.*, 1974), making this a most useful species for future biochemi-cal and genetic studies.

III. LIFE CYCLE

Dictyostelium discoideum Raper (1935) has been studied most inten-sively in the laboratory. Its life cycle is diagrammed in Fig. 76 and its sorogenesis in Fig. 77. When the elliptical or bean-shaped spore germinates, a single amoeboid cell emerges through a longitudinal split in the spore wall. The trophic stage consists of uninucleate amoeboid cells with filose pseudopodia (Fig. 78). Lobose pseudopodia may also be produced. The nucleus is less conspicuous than that of protostelids and myxomycetes since it has usually from two to five small peripheral nucleoli instead of one larger, central nucleolus (Fig. 79). However,

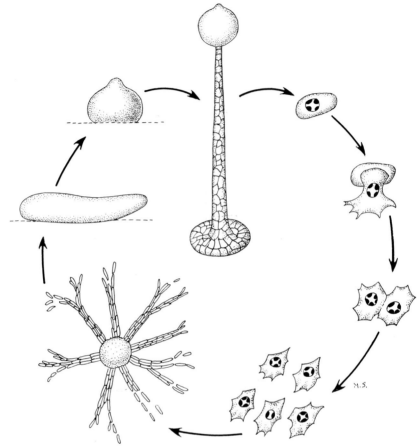

Fig. 76.   Diagram of life cycle of *Dictyostelium discoideum*. From Olive (1970).

in a culture of *D. mucoroides* stored in a 16°C incubator, we found that most nuclei had a single peripheral nucleolus, whereas some had two and a very few contained three. Proteins with basic properties similar to the histones of *Physarum*, the euprotistan *Tetrahymena*, and calf thymus are present in the nucleus. A single contractile vacuole and generally several food vacuoles that contain ingested bacterial cells in various stages of digestion occur in the cell. It is also not uncommon to find within an amoeba engulfed cells of the same dictyostelid—a situation referred to as "cannibalism." The cells divide in the manner described for protostelids, the generation time of the amoebae being about 3 hours. Following nuclear division, the cell constricts around the middle and the two daughter cells quickly separate. The haploid chromosome number is seven in *D. discoideum* (Fig. 80) and eight

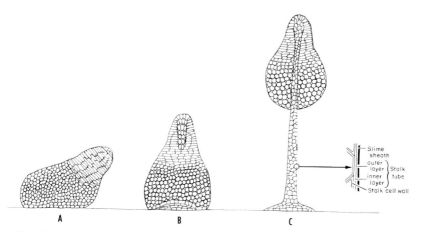

Fig. 77. Diagram of sorogenesis in *Dictyostelium discoideum:* (A) pseudoplasmodium at conclusion of migration; (B) beginning of stalk development; (C) later stage in stalk formation. Note upper prestalk area and lower prespore area.

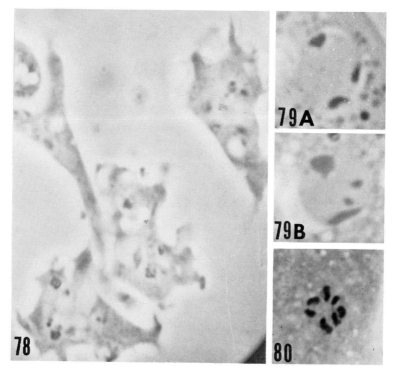

Figs. 78–80. *Dictyostelium.* Fig. 78. Amoebae of *D. purpureum* showing filose pseudopodia, ×1700. Fig. 79, A and B. Nuclei of *D. irregularis* showing peripheral nucleoli. Phase contrast ×2700. Fig. 80. Late prophase in *D. discoideum;* the seven chromosomes are apparent, ×2660. Aceto-orcein stain. Fig. 80 from Wilson (1953).

or nine in *Polysphondylium violaceum* (Wilson and Ross, 1957). Diploid and aneuploid spores and amoebae have also been detected, and some strains of *D. discoideum* have proved to be relatively stable diploids (Ross, 1960; Sussman and Sussman, 1962). These differences in ploidy, which apparently do not stem from a true sexual process, have other probable explanations that are noted in Sections VI,B and VIII. Cells with more than one nucleus are sometimes found, and these divide by plasmotomy.

When the amoebae have reached a certain density of population and have virtually exhausted the available food supply, usually within a few days after the culture is started, they cease feeding, gradually lose their food vacuoles, and stop dividing. The period during which these changes occur is known as "preaggregation." It precedes aggregation and lasts for about 4–8 hours. Toward the end of this period, small groups of cells become associated and begin to attract other cells toward them, until streams of cells are moving into aggregation centers. The attractant, first called "acrasin" (Bonner, 1947), has been found to be cyclic 3',5'-adenosine monophosphate (cyclic AMP) (Konijn *et al.*, 1967, 1969). Only sensitized amoebae and not trophic amoebae are attracted strongly by the substance. At the end of aggregation a thin, protective slime sheath is secreted around the cell mass, which is now known as the pseudoplasmodium or "slug." In *D. discoideum* the slug migrates across the substrate, leaving behind a slime track derived from the collapsed slime sheath. The period of migration varies with environmental conditions, especially humidity. When migration ceases, the pseudoplasmodium sits upright and begins to form the sorocarp (Fig. 77, A and B).

The sorogen now begins the remarkable process of forming a multicellular stalk and a terminal sorus of spores, a process requiring about 6 hours (Fig. 77C). A group of rearguard cells at the base remain on the substrate and form the basal supporting disc of the stalk. The remainder of the sorogen consists of an anterior region of prestalk cells and a larger posterior region of prespore cells, which are morphologically and physiologically distinct areas that differentiate during aggregation and migration. The first stalk cells appear in the papilla-like upper part of the sorogen and can be recognized by their conspicuous vacuolation. They pass down through the center of the sorogen, with others following in succession until they join the basal disc. In the process they become more highly vacuolated, secrete cellulose walls, and finally become empty, dead cells. During stalk development a double-layered cellulose tube is produced around the core of closely packed stalk cells. Stalk tube and stalk cell walls, along with the slime sheath, perform

the supportive function of the stalk. The outer layer of the stalk tube
is probably homologous with the stalk tube of protostelids and *Acytostelium* (Gezelius, 1959; Olive, 1970). We therefore prefer the term
"stalk tube" for the double-layered structure in the Dictyosteliidae rather
than the previously used "stalk sheath."

As the stalk continues to elongate, the remainder of the sorogen rises
upon it, probably by amoeboid movement of cells in contact with the
stalk tube and the slime sheath surrounding the sorogen. When the
prestalk cells are all used up, only a mass of developing spores is left.
Encystment occurs without a preceding mitosis or cytokinesis, and the
mature spores are encased in cellulose walls. The spores are embedded
in a mucilaginous matrix that probably increases their disseminability
by insects and mites. The stages of morphogenesis are discussed in
greater detail in Sections VI and VII of this chapter.

There has been much uncertainty as to whether a true sexual process
occurs in dictyostelids. Reports that syngamy, karyogamy, and meiosis
are associated with fruiting (Wilson, 1953; Wilson and Ross, 1957) have
not been substantiated. Recent crossing experiments and ultrastructural
studies have yielded evidence indicating that a sexual process is involved
in macrocyst development. Macrocysts develop from cell aggregates
under conditions that discourage fruiting. They are known to occur
in isolates of at least five species of dictyostelids, including *D. discoideum*. Their ontogeny is discussed in Sections VI,E and VII,B.

IV. KEY TO FAMILIES AND GENERA

A key to the two families and four genera of dictyostelid cellular
slime molds is given below. *Coenonia*, which is known only from the type
collection, is tentatively assigned to this subclass.

Sorocarps with very slender, hollow stalk tube .... **Acytosteliidae**
                                Single genus: *Acytostelium*
Sorocarps with a broader stalk tube that is filled with a framework of empty cells .... **Dictyosteliidae**
    Apex of sorocarp a globose sorus of spores, not cupulate
        Sorocarps typically with terminal sorus only but some-
        times having, in addition, irregularly distributed secondary
        branches or sessile sori along the stalk .... *Dictyostelium*
        Sorocarps with large terminal sorus and regular whorls of
        smaller stalked sori along the main stalk .... *Polysphondylium*
    Apex of sorocarp expanded into a denticulate spore-bearing
    cupule .... *Coenonia*

V. KEYS TO SPECIES AND DESCRIPTIONS OF TAXA

In the following discussions of dictyostelid species, our own unpublished distributional data are added to those previously published (mostly by E. W. Olive, Raper, and Cavender).

A. Acytosteliidae Raper (Raper and Quinlan, 1958)

Acytostelium Raper, the sole genus of the family, contains three species, all of which have delicate, relatively inconspicuous sorocarps, each consisting of a narrow hollow stalk, little or no broader than those of protostelids, bearing a terminal sorus of spores.

Key to the Species of *Acytostelium*

Spores globose; regular in shape .... *A. leptosomum*
                                                        Raper (Raper and Quinlan, 1958)
Spores not as above
  Spores typically subglobose to
  ovoid and somewhat irregular .... *A. irregularosporum*
                                                        Hagiwara (1971)
  Spores ellipitical .... *A. ellipticum*
                                                        Cavender (1970)

Acytostelium leptosomum has stalks that are up to 1.5 cm tall and round spores that measure 5–7 μ in diameter. It has been found in forest soils, humus, insect frass, and on a decaying agaric, in the United States, Brazil, Spain, and East Africa. The sorocarps of *A. ellipticum* are smaller and more delicate, measuring 0.2–1 mm tall and bearing elliptical spores that measure 2–3 × 5.5–8 μ. It has been reported only from forest soil collections in Trinidad and Colombia. *Acytostelium irregularosporum* has sorocarps of about the same size as *A. ellipticum* and somewhat irregular subglobose to ovoid spores 3.8–5.8 × 4.8–8 μ in size. It has been isolated from leaf mold in Japan.

The morphogenesis of the better known *A. leptosomum* has been studied by Raper and Quinlan (1958). The amoebae resemble those of *Dictyostelium*, but the nuclei contain fewer peripheral nucleoli, usually from one to three (Olive, 1970). The species has been grown in culture primarily on bacteria, but we have found that it also grows and fruits on yeasts in the absence of bacteria. Encysted amoebae, or microcysts, are spherical and therefore similar in appearance to the spores.

Aggregation in *A. leptosomum* is by streaming of amoebae. When the aggregate is complete it remains flattened, and usually several papilla-like masses of amoebae appear on its surface and proceed to develop directly into mature sorocarps (Fig. 81). The smallest aggregates pro-

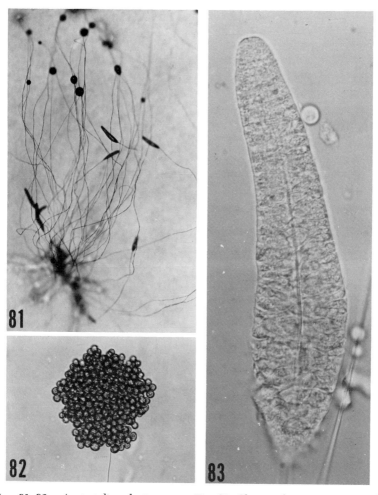

Figs. 81–83. *Acytostelium leptosomum.* Fig. 81. Cluster of sorocarps arising from a single pseudoplasmodium, ×40. Fig. 82. Sorus of round spores at the tip of the narrow stalk, ×350. Fig. 83. Sorogen showing developing stalk inside and transverse orientation of cells, ×825. From Raper and Quinlan (1958).

duce only one fruiting body. Throughout sorogenesis a thin slime sheath covers the entire sorocarp. Whether all cells in the somewhat spindle-shaped sorogen take part in stalk formation is unknown. In vertical sections of the sorogen, the cells appear elongated with their long axes at right angles to the stalk (Fig. 83). In transverse section they resemble pie-shaped wedges. The tip of the sorogen is occupied by one or two "cap cells." The stalk is only about 1–2 μ in diameter, as in protostelids, but here there is no information on how its narrow lumen is maintained during sorogenesis. It would be interesting to determine whether there is a narrow extension of the cap cell into the upper end of the developing tube that resembles the extension of the steliogen into the stalk tube of a protostelid.

When culmination ceases, the sorogen rounds out and all of its cells encyst to form the globose spores (Fig. 82). By the time the sorus matures, the thin surrounding slime sheath has disappeared, leaving only a mucous mass of spores. The delicate sorophores often bend or collapse, bringing the spores to the substrate surface. The standing sorocarps appear ideally suited for spore distribution by mites and small insects.

Spore and microcyst walls, as well as stalks, of *Acytostelium* have been shown to be chiefly cellulosic by means of X-ray diffraction, chemical analysis, and electron microscopy (Gezelius, 1959). Since *Acytostelium* sorogens have only one type of cell and all cells become spores, the genus should have some unique advantages in the analysis of biochemical events accompanying morphogenesis.

An interesting observation made by Dr. Brian Shaffer (personal communication) is that fruiting bodies of *A. leptosomum* that develop on nonnutrient, washed agar in the absence of a food organism tend to be quite small and sometimes consist only of a hairlike stalk bearing a single terminal spore. We have repeated these conditions and obtained fruiting bodies with as few as two spores. These minute structures are indistinguishable from the sporocarps of protostelids, and their occurrence lends support to the theory that dictyostelid cellular slime molds have evolved from protostelids by way of *Acytostelium* (Olive, 1970).

B. Dictyosteliidae Rostafinski (1875)

1. *Dictyostelium* Brefeld (1869)

This is the largest genus of cellular slime molds. The 19 or so known species are found in forest soils and humus and on rotting wood, moribund plant structures, decaying fungi, and dung of a variety of animals. Several are worldwide in distribution.

Key to the Species of *Dictyostelium*

1. Spores globose
  2. Sorocarps minute, with single terminal sorus .... **D. lacteum**
                                                         van Tieghem (1880)
  2. Sorocarps large, occasionally branched, with terminal sorus and lateral sessile ones .... **D. rosarium**
                                                Raper & Cavender (1968)
1. Spores elliptical, bean-shaped, or reniform
  3. Sorocarps without cramponlike bases
    4. Pseudoplasmodia or sorogens migrating over substrate
      5. Horizontal stalk usually deposited on substrate during migration
        6. Sori purple .... **D. purpureum**
                                                E. W. Olive (1901)
        6. Sori white or yellowish
          7. Sorocarps irregular, simple to coremiform .... **D. irregularis**
                    Olive, Nelson & Stoian. (Nelson *et al.*, 1967)
          7. Sorocarps regular, never coremiform
            8. Stalks typically less than 1.5 cm long .... **D. mucoroides**
                                           Brefeld (1869)
            8. Stalks mostly 1.7–2.5 cm long .... **D. giganteum**
                                           Singh (1947)
      5. Stalk typically not deposited on substrate during migration
      9. Pseudoplasmodium sluglike, producing a single sorocarp
        10. Stalk with basal disc; spores mostly 2.5–3.5 × 6–9 μ **D. discoideum**
                                           Raper (1935)
        10. Stalk without basal disc; spores 2.5–3.5 (−5) × 7–12 (−26) μ; horizontal stalk sometimes produced on substrate .... **D. dimigraformum**
                                           Cavender (1970)
      9. Pseudoplasmodium extremely elongated, giving rise to a coremiform cluster of sorocarps .... **D. polycephalum**
                                           Raper (1956)
    4. Migrating stage absent, sorocarps relatively small
      12. Sorocarps frequently with small lateral branches; spores 3–3.5 × 5–7 μ .... **D. minutum**
                                          Raper (1941b)
      12. Sorocarps rarely branched; spores narrow, 1.5–2 × 3.5–5 μ .... **D. deminutivum**
                                      Anderson *et al.* (1968)
  3. Sorocarps with branched cramponlike bases, mostly 1.5–5 mm tall
    13. Both terminal and lateral sessile sori present .... **D. laterosorum**
                                     Cavender (1970)

13. Terminal sorus present; lateral sessile sori absent
    14. Sori white, yellowish, or grayish to brownish olive;
        cramponlike bases usually well developed
        15. Sori yellowish to brownish olive; stalks relatively
            bulky and often long, yellowish or brownish purple .. *D. rhizopodium*
                                                    Raper & Fennell (1967)
        15. Sori white or cream colored; stalks usually compara-
            tively short, often strongly tapered, bluish purple ... *D. coeruleo-stipes*
                                                    Raper & Fennell (1967)
    14. Sori lavender to purple or dark smoky; crampon bases
        distinct or poorly developed
        16. Sori some shade of grayish lavender to vinaceous; stalks
            comparatively long and thin, purplish; crampons fairly
            well developed .... *D. lavandulum*
                                                    Raper & Fennell (1967)
        16. Sori darker, deep vinaceous gray to dark smoky; stalks
            typically thin and delicate; crampons small and digitate
            or bases consisting merely of irregularly grouped
            vacuolate cells .... *D. vinaceo-fuscum*
                                                    Raper & Fennell (1967)

K. B. Raper (personal communication) believes that *D. aureum* E. W. Olive (1901), with yellow pseudoplasmodia and sorocarps; *D. sphaerocephalum* (Oud.) Sacc. and March. (see E. W. Olive, 1902), with coarse stalks and large sori; *D. roseum* van Tieghem (1880), with rose-colored sori; and *D. brevicaule* E. W. Olive (1901), with short erect sorocarps and white sori, may also be valid species. However, an adequate judgment cannot be made with the scant information available on them at this time.

A number of *Dictyostelium* species are known from both temperate and tropical zones, but *D. laterosorum, D. dimigraformum,* and the four crampon-based species of Raper and Fennell have been found only in the tropics and subtropics. Since the latter areas have been searched less diligently for dictyostelids, more new species are likely to be found there in the future.

Dictyostelium mucoroides (Figs. 84B and 85) was not only the first described species but is also the most commonly encountered one, having been found around the world and from Alaska to the tropics. It is one of several dictyostelids that lay down a horizontal stalk on the substrate during migration of the elongate sorogen. It has been studied by a number of investigators (see Bonner, 1967). Aggregation is characterized by long conspicuous streams of amoebae that build up a relatively large aggregation center. The mature aggregate becomes vertically elongate, at which time the stalk begins to appear within it. Shortly thereafter, the entire mass assumes a horizontal position on the substrate and the

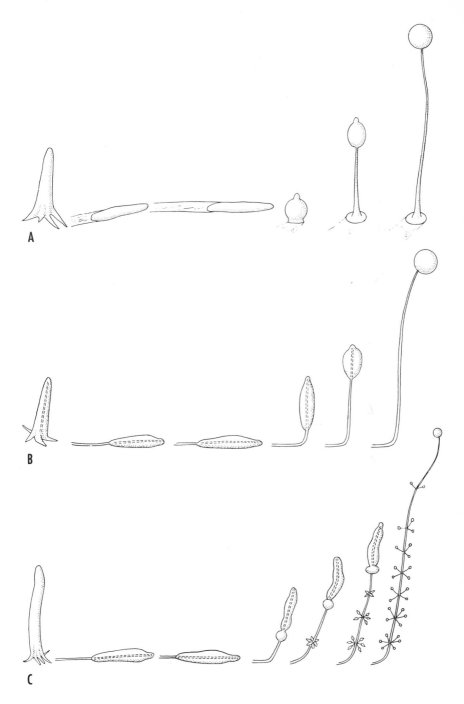

Fig. 84. Sorogenesis in dictyostelids: (A) *Dictyostelium discoideum;* (B) *D. mucoroides* and *D. purpureum;* (C) *Polysphondylium.* Redrawn from Bonner (1957).

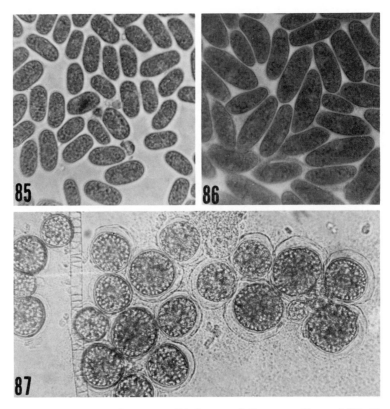

Figs. 85–87. *Dictyostelium.* Fig. 85. Spores of *D. mucoroides,* ×1400. Fig. 86. Spores of *D. discoideum,* ×1400. Fig. 87. Macrocysts of *D. mucoroides,* ×210. Figs. 85 and 86 from Raper (1951); Fig. 87 from Blaskovics and Raper (1957).

sorogen of amoeboid cells migrates across its surface, laying down a continuous horizontal stalk posteriorly as it proceeds (Fig. 84B). The period of migration varies with environmental conditions but eventually the sorogen becomes vertically directed and culminates to form the aerial part of the stalk and, finally, a relatively large terminal sorus with short-cylindrical to bean-shaped spores that commonly measure 2.5–4.5 × 4–9 μ (Fig. 85). During migration and culmination the developing part of the stalk can be seen throughout the length of the sorogen.

Excessive horizontal stalk development may occur in culture, especially under conditions of high humidity, and is probably indicative of abnormal cultural conditions. However, a certain amount of horizontal stalk helps support the aerial part of the sorophore, and the migratory phase of this and other species with or without horizontal stalk formation favors movement of pseudoplasmodia and sorogens to more favorable spots for fruiting and spore dispersal.

Dictyostelium mucoroides, like most other mycetozoans, may form uninucleate microcysts by the rounding up and encystment of individual amoebae. In addition, some isolates produce macrocysts (Fig. 87).

In typical isolates of *D. mucoroides* it is not uncommon during culmination for an occasional mass of cells to be constricted off from the lower part of the sorogen and remain behind on the stalk. Such a mass gives rise to a small side branch with a small sorus of spores. One unusual variant of *D. mucoroides* collected in Costa Rica (Huffman and Olive, 1963) produces aggregates that usually give rise to several sorogens. As each sorocarp develops, it is common for an extensive cell mass from the rising sorogen to be left behind for a considerable distance along the midregion of the stalk. Each such mass becomes subdivided into secondary sorogens that produce as many small lateral branches, sometimes exceeding two dozen in number. The variant appears to show relationships with *D. polycephalum* as well as *Polysphondylium*, and Huffman and Olive have suggested that *D. mucoroides*, which has numerous variants, has occupied a plastic and pivotal position in the evolution of dictyostelid cellular slime molds. However, an analysis of DNA base ratios in dictyostelids by Dutta and Mandel (1972) indicates a lack of any close relationship between *D. mucoroides* and *Polysphondylium*.

Dictyostelium mucoroides var. *stolonifera* Cavender and Raper (1968) differs from typical collections in having spores that can germinate immediately on the substrate, after which the amoebae can aggregate without multiplication to form a number of small sorocarps *in situ*.

Dictyostelium giganteum, isolated from a compost heap in England, has sorocarps that are commonly 1.7–2.5 cm, occasionally up to 3 cm, in length, frequently with one or more lateral branches. The spores resemble those of *D. mucoroides* and measure 2–3.5 × 5–10 μ. It has recently been collected on several occasions in Wisconsin (K. B. Raper, personal communication).

Dictyostelium discoideum (Figs. 84A, 86, 88, 100, 102, and 103), although collected less frequently than *D. mucoroides*, is much better known than any other dictyostelid because it has been the subject of intensive investigation in a number of laboratories. Its life cycle has been diagrammed (Fig. 76) and briefly discussed. Only recently have macrocysts been found in this species. Because it is the major subject of the detailed discussions that follow in Section VI, it will not be further described here. It has been isolated from forest soils and humus, rotting wood, decayng fungi, dead stems, and skunk dung in the United States, Canada, Mexico, and Brazil.

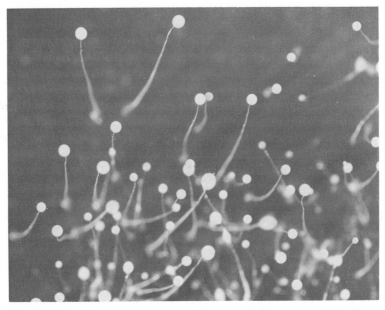

Fig. 88. *Dictyostelium discoideum* sorocarps, ×12. Courtesy of K. B. Raper.

Dictyostelium minutum, a fairly common species, has small sorocarps, mostly 500–850 μ tall, resembling the ones produced by *D. mucoroides.* Apparently, there is no preaggregation period and a conspicuous radiate aggregation pattern is generally absent, as is a migration phase. The stalk usually consists of a single tier of cells and small lateral branches are not infrequent. Macrocysts have been observed in some isolates. The species has been found in forest soils, rotting wood, insect frass, and decaying fungi in the United States (including Alaska), Canada, Mexico, Ecuador, New Zealand, Tahiti, Japan, and the Seychelles.

Dictyostelium purpureum is a rather common species that resembles *D. mucoroides* but may readily be distinguished from the latter by its pigmented spores, which cause the sori to appear dark purple. The pigment is located in the spore wall. A stalk is deposited on the substrate during migration (Figs. 84B and 101), and the spores are similar in size and shape to those of *D. mucoroides.* Macrocysts are produced by some isolates. It has been isolated from forest soils, dead vegetation, tree bark, and the dung of various animals in the United States (including Hawaii), Canada, Mexico, Malaya, Fiji, Japan, New Guinea, and East Africa.

Dictyostelium lacteum is one of the two species of *Dictyostelium* known to have globose spores. Aggregation is by streaming. Unless the aggregate is quite small it segments to form several sorogens, each of which gives rise directly to a small sorocarp with slender stalk that consists primarily of a single tier of cells. The tip cell is often long, narrow, and tapered and the stalk tip occasionally has a very narrow apical, acellular extension resembling the stalk of *Acytostelium,* which along with the round spores suggests a relationship with *A. leptosomum* (Bonner, 1967). A tentative method of classification based on relative complexity of developmental control systems and morphogenesis proposed by Robertson and Cohen (1972) also indicates a primitive position for this species in the genus. It occurs in forest soils and on rotting wood, decaying agarics, and frog dung in the United States, Canada, and Europe.

Dictyostelium deminutivum has the smallest fruiting bodies of any known member of the genus. They are mostly 350–600 μ tall, and the spores are reniform to capsule-shaped and quite narrow. Aggregation is not by streaming but by the movement of amoebae singly or in small groups into the centers. The species was isolated originally from leaf mold collected in a humid tropical forest in Mexico and later from the Cascade Mountains of Washington (Mishou and Haskins, 1971).

Dictyostelium polycephalum is a unique and interesting species. In agar cultures broad sheets of amoebae move into the aggregation centers, which then divide into a variable number of pseudoplasmodia that migrate across the substrate. The latter are unusually long and narrow, sometimes reaching a centimeter in length (Fig. 89). At the end of migration the slug assumes a globular shape and usually becomes subdivided into several parts that give rise to as many sorocarps, the stalks of which adhere along much of their length in coremiform fashion to form a compound fruiting structure (Figs. 90 and 91). Fruiting does not occur readily in the laboratory but may be induced by growing the organism in three-membered culture with *Aerobacter aerogenes* and the fungus *Dematium nigrum* at 30°C (Raper, 1956). K. B. Raper (personal communication) has also found that activated charcoal added to the agar surface stimulates sorogenesis. It is of interest to note that the migrating slug of this species shows no differentiation into prespore and prestalk areas, which is consistent with the need for all cells to remain relatively unmodified until the pseudoplasmodium subdivides into the papillae that produce the sorocarps. Although *D. polycephalum* is not a common species it has a wide distribution, having been found

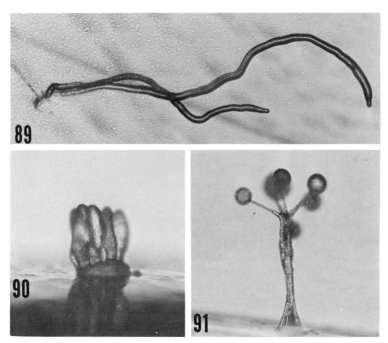

Figs. 89–91. *Dictyostelium polycephalum.* Fig. 89. Pseudoplasmodia, ×25. Fig. 90. Sorogenesis, ×75. Fig. 91. Compound coremiform fructification comprised of seven sorocarps, ×70. From Raper (1956).

in forest soil and humus in the United States (including Hawaii), Canada, East Africa, and the Seychelles.

Dictyostelium rosarium has a sorocarp with large terminal sorus and numerous smaller, sessile sori along the stalk, giving it a beaded appearance (Fig. 92). The spores are round. The only other species with beaded sorocarps is the crampon-based *D. laterosorum* found in tropical America. Its sorocarps are readily distinguished from those of *D. rosarium* by their bluish gray to light violet color, the divided stalk base, and the elongated spores. The lateral sori of these two species may indicate a relationship with *Polysphondylium. Dictyostelium rosarium* is an uncommon species that has been isolated from semiarid mesquite forest soil and from the dung of horse and deer in the United States and Mexico.

There are five known crampon-based species of *Dictyostelium*: *D. laterosorum* and the four species *D. rhizopodium, D. lavandulum, D. coeruleo-stipes,* and *D. vinaceo-fuscum* described by Raper and Fennell (1967) from forest soils and humus in the southern United States and

Latin America. The stalk bases are subdivided in a characteristic fashion (Fig. 93). The sori or sorophores, or both, are yellowish, brown, violet, or purplish in color.

The recent publications of Hagiwara (1971, 1972, 1973a,b) have extended the known range of several dictyostelids and described a number of new species. Unfortunately, sufficient information on morphogenesis and the presence or absence of a migrating stage in the new species is generally not supplied, making it difficult to include them in the key. Hagiwara (1971, 1973a) isolated from soil, humus, and dung in Japan and New Guinea a species that he considered to be *D. sphaerocephalum* (Oud.) Sacc. and March. It produces white sorocarps 0.1–8 mm tall with sori 20–360 μ in diameter and spores measuring 2.8–5.6 × 4.8–9.2 μ. One isolate was found to produce macrocysts. *Dictyostelium firmibasis* Hagiwara (1971), from leaf mold and a decaying bracket fungus, forms white to yellowish white sorocarps 1.3–9 mm tall, with sori 110–400 μ in diameter and spores that measure 2.8–3.8 × 6.6–9.2 μ. Differences between this species and the foregoing seem slight. *Dictyostelium delicatum* Hagiwara (1971), isolated from leaf mold in Japan, has white sorocarps 2–4.6 mm tall, sori 70–120 μ in diameter, and spores measuring 2.6–3.8 × 5.2–8.4 μ. *Dictyostelium monochasioides* Hagiwara (1973a), from soil and leaf mold in New Guinea, has white sorocarps

Fig. 92. *Dictyostelium rosarium;* sorocarps showing terminal sorus and smaller lateral, sessile sori, ×20. From Raper and Cavender (1968).

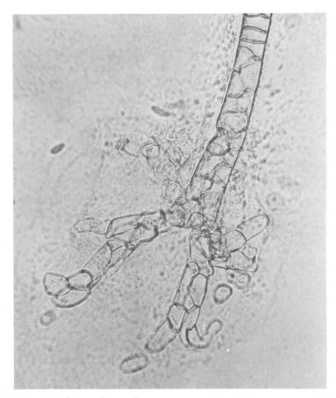

Fig. 93. *Dictyostelium rhizopodium;* crampon base of sorocarp, ×450. From Raper and Fennell (1967).

that often show major branching near the base. They are 0.2–3.1 mm tall, with sori 30–210 μ in diameter and spores 2.8–4.6 × 4.4–9.6 μ. A satisfactory evaluation of these proposed new species and their relationships to other members of the genus can be made only after more information on their development becomes available.

2. *Polysphondylium* Brefeld (1884)

This genus contains two widespread and rather common species that are among the most attractive of the cellular slime molds. Their mode of development offers another example of the remarkable diversity of this group. Aggregates commonly form by conspicuous amoebal streaming. As in *Dictyostelium mucoroides,* the sorogens, which are often quite large, frequently migrate and leave behind a continuous stalk on the substrate (Fig. 84C). As the sorogen becomes vertically oriented and begins forming the long aerial portion of the multicellular stalk, it intermittently segments posteriorly to produce a doughnut-shaped mass of amoebae around the stalk. These doughnut-shaped masses differentiate sequentially, beginning with the lowermost one, by subdividing into several small secondary sorogens that develop into as many secondary stalked sori (Figs. 94–96). This results in a series of whorled small branches along the stalk, and the remaining terminal mass at the tip of the main stalk becomes a single large sorus. The sorocarps are often quite long and tend to collapse onto the substrate.

As in the migrating slugs of *D. polycephalum,* there is no differentiation into prespore and prestalk regions within the migrating or culminating sorogens of *Polysphondylium.* The maintenance of this "embryonic" state of the cells is undoubtedly related to the role of the sorogen in cutting off successive doughnutlike segments of similarly unspecialized cells, which then give rise to secondary stalked sori. Sometimes small aggregates in *Polysphondylium* produce small sorocarps with only a terminal sorus and no secondary branches. These resemble the sorocarps of *Dictyostelium.* The two commonly recognized species of *Polysphondylium* may readily be distinguished on the basis of spore color.

Polysphondylium violaceum Brefeld (1884) generally has larger sorocarps (up to 2 cm or more in length) and violaceous sori. Some isolates produce macrocysts. Rai and Tewari (1963) have described a variant form of *P. violaceum* with lateral branches that are more often irregularly arranged than in whorls. They concur with the opinion of Huffman and Olive (1963) that the species has probably evolved from the *Dictyostelium mucoroides* complex. *Polysphondylium violaceum* has been isolated from cultivated and forest soils, humus, rotting wood, dead vegetation, and the dung of various animals in the United States (including Hawaii), Canada, Central America, Ecuador, Europe, New Zealand, Fiji, Japan, India, and East Africa.

Polysphondylium pallidum E. W. Olive (1901) has somewhat more delicate sorophores and white sori (Figs. 94–96). It has been isolated from forest soils and humus, decaying basidiocarps, and various kinds of dung in the United States (including Hawaii), Canada, Latin America, Tahiti, Samoa, Fiji, New Zealand, New Guinea, Japan, Malaya, Ceylon, East Africa, and the Seychelles.

Polysphondylium album E. W. Olive (1901) is generally considered only a variant form of *P. pallidum. Polysphondylium candidum* Hagiwara (1973b), isolated from soil and humus in Japan and described as having white sorocarps 0.3–14.3 mm long, a terminal sorus 40–155 μ in diameter,

Figs. 94–96. *Polysphondylium pallidum.* Fig. 94. Sorogenesis, ×28. Fig. 95. Mature sorocarps by reflected light, ×15. Fig. 96. Single whorl of secondary sori, ×125. From Raper (1960).

one to six whorls of secondary branches, and oblong–ellipsoid spores 2.8–7.4 × 5.2–13.2 μ, is distinguishable from *P. pallidum* primarily on the basis of spore size. Because spore size is subject to alteration with changes in ploidy, more information is needed before this species can be accepted as valid.

3. Coenonia van Tieghem (1884)

The single species, *Coenonia denticulata,* has not been reported since van Tieghem originally discovered it on decaying beans in France. It is certainly one of the most unusual of the cellular slime molds. There are no published illustrations of it, but it is described as having sorocarps 2–3 mm tall, a branched base expanded into a cramponlike structure, and a cupulate apex bearing a yellowish sorus of globose spores. Each peripheral cell of the stalk and cup is said to bear a single tooth or papilla. Some of the sorocarps are further characterized by having a verticel of three smaller branches similar to the main axis. When van Tieghem transferred a mass of amoeboid cells from a sorogen to a nutrient droplet, the cells produced a new and smaller sorocarp.

Coenonia denticulata shows some evidence of relationship with the crampon-based species of *Dictyostelium* as well as *Polysphondylium,* but too little is known about it to speculate much further about its position among the cellular slime molds. It is a unique species that should be reisolated and studied in detail.

VI. DETAILS OF THE LIFE CYCLE

The literature on morphogenesis in dictyostelids has recently grown exponentially as more investigators discover their usefulness and ease of handling in the laboratory. Some of the major aspects and the significance of these studies are considered in this section. Because investigations have centered primarily on *Dictyostelium discoideum,* it may be assumed throughout the following account, unless otherwise stated, that this is the species under discussion.

A. Preaggregation

During the trophic stage the amoebae wander about haphazardly, feeding on bacteria, dividing by binary fission, and showing no evidence of mutual attraction. Our observations (consistent with those of Robertson and Cohen, 1972) on amoebae in "sandwich" preparations reveal that the advancing front of the amoeba is hyaloplasmic. Lobose extensions appear in succession along the front, usually (but not always) followed quickly by the appearance of one to several filose pseudopodia from each. Continued movement in the same direction is followed by a lobose expansion that overtakes and includes the filose ones, and the process is repeated. The posterior end of the amoeba appears to be somewhat sticky. An amoeba may change direction at any time by putting out pseudopods from any area of the cell. The cells are largest during the trophic stage and show a decrease in size as they approach the aggregation stage (Bonner and Frascella, 1953).

There is a period of about 4–8 hours prior to aggregation known as "preaggregation" (also called "interphase"), during which the physiology and structure of the amoebae alter conspicuously. The cells stop feeding, discharge the contents of their food vacuoles, become smaller in size, and show a change in staining properties. Metachromatic granules, stainable with Toluidine Blue, appear in the cytoplasm. With cessation of feeding and discharge of food vacuoles, the amoebae must now depend on their endogenous food reserves. Centrifugation of trophic amoebae of *D. mucoroides* and *D. purpureum* to rid them of bacteria may bring about immediate aggregation without an intervening preaggregation period.

During preaggregation the amoebae reach a point at which their rate of movement doubles (Takeuchi, 1960; Samuel, 1961), after which they become more elongate and enter into aggregation streams. Cell division probably does not occur during late preaggregation or aggregation.

Using a temperature-sensitive axenic strain (Ax-3), Katz and Bourguignon (1974) were able to synchronize cell populations. The lengths of the various stages in the cell cycle were determined as follows: G_1 (between mitosis and DNA synthesis), 2 hours; S (DNA synthesis), 1–2 hours; G_2 (between DNA synthesis and mitosis), 4 hours; M (mitosis), 1–2 hours. Aggregating cells were found to be at mid-G_2 phase. The limiting factor in the stimulation of aggregation was found to be starvation. Synchronized cells starved at various stages of the cell cycle required the same length of time for aggregation.

B. Aggregation

Aggregation may first be detected at the end of preaggregation by the appearance of small groups of amoebae that begin to attract other amoebae. These thus build up centers that attract many more amoebae, which eventually converge in more or less radiate streams (Figs. 97–99). Ennis and Sussman (1958) propose that the aggregates are started by special "initiator cells" (I cells) occurring in a fixed proportion, estimated to be 1/2200, in a population. Sackin and Ashworth (1969) and Ashworth and Sackin (1969) have also concluded that I cells are present and that the most effective ones are those characterized by a low level

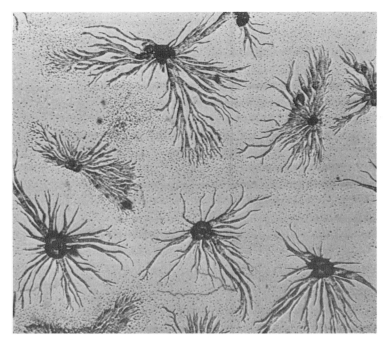

Fig. 97. *Dictyostelium discoideum;* aggregation by amoebal streaming, ×13. Courtesy of K. B. Raper.

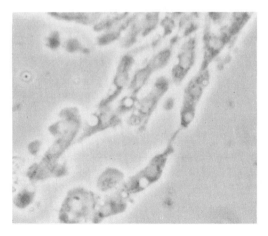

Fig. 98. *Dictyostelium discoideum;* amoebae from an aggregate. Note slender extension between two cells on right. Phase contrast ×975. From Gregg (1964).

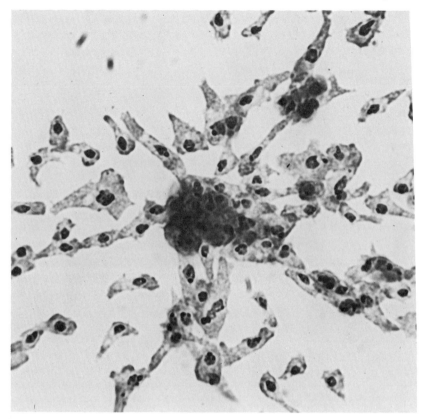

Fig. 99. *Dictyostelium discoideum;* aggregating cells stained with silver nitrate. Note single nucleus in each cell, ×667. From Bonner (1947).

of aneuploidy, especially those with $n + 1$ chromosomes carrying a double dose of genes influencing aggregation. They maintained that frequencies of diploid, haploid, and aneuploid spores can be determined by volumetric measurements and arrived at an estimated frequency of I cells close to that suggested by Ennis and Sussman. Supporting evidence that gradations in ploidy can be determined with such precision by these methods is needed.

Gerisch (1961), Shaffer (1962), Bonner (1967), and others have explained why they do not think that I cells occurring in a fixed proportion are present in *D. discoideum.* Serious doubts are also raised in experiments by Konijn and Raper (1961), who have plated out separate droplets containing about 100 amoebae each and, in the majority of cases, have obtained aggregation. Therefore, if present, I cells would have to be far more frequent than claimed by Ennis and Sussman.

Whereas aggregation can first be accurately identified in *D. discoideum* by the appearance of a small group of amoebae, it is possible in *D. minutum* (Gerisch, 1963) and *Polysphondylium violaceum* (Shaffer, 1961) to identify individual "founder cells" around which aggregation begins. These cells, which do not occur in a fixed proportion in a population, become rounded and sedentary, after which other cells in the vicinity begin moving toward them.

It has been known for some time that amoebae during preaggregation and aggregation exhibit a surface adhesiveness or "stickiness" and tend to adhere to each other, somewhat briefly during preaggregation but for longer periods during aggregation. When the cells move apart, fine connecting strands frequently appear between them, draw out, and pull away or snap in two (Fig. 98). Adhesiveness of these amoebae to each other and to inorganic surfaces is low during the trophic stage but increases rapidly through preaggregation and stages of morphogenesis (Yabuno, 1971).

With the aid of immunological techniques, Gregg (1961, 1966), Riedel and Gerisch (1969), and others have shown that a variety of antigens (26 detected) appear at the cell surface in the period extending from trophic stage to sorocarp maturation, and that a particular combination of antigens characterizes any specific stage. Certain antigens typical of early stages of development are replaced by others during later stages. Adhesiveness of cells is considered to be associated with aggregation antigens, a view that is supported by the fact that aggregation is inhibited by antibodies to these antigens. Species that are able to coaggregate with *D. discoideum* accumulate a serologically detectable substance that cross-reacts with the *D. discoideum* antigens, whereas those that are unable to coaggregate do not.

A number of important publications on the nature of cell adhesiveness have appeared recently. The marked increase in cell cohesiveness that accompanies the approach of aggregation is closely correlated with the appearance of a proteinaceous agglutination factor with a specificity that enables it to bind with monosaccharides having a galactose configuration (Rosen *et al.*, 1973). Cell-to-cell adhesion is controlled by two classes of contact sites that occupy only a small proportion of the cell surface (Beug *et al.*, 1973a,b). Sites A, which are responsible for end-to-end contacts, increase from a near-zero level during the trophic stage to a high level just before aggregation; sites B, which are responsible for adhesion over the whole cell surface, are essentially constant in frequency throughout trophic and developmental stages. Sites A are more important in the development of aggregation streams, although sites B also contribute by promoting lateral contacts between cells in the streams. Specific antibodies are able to block the effects of one kind of site without affecting the other. A mutant of *D. discoideum,* known as *aggregateless-50,* which fails to form aggregation streams, lacks A sites. These studies obviously have considerable significance for the overall understanding of animal cell behavior during morphogenesis.

At least some of the interconnecting strands observed between amoebae during preaggregation and early aggregation have been found to be temporary protoplasmic bridges or anastomoses that permit an exchange of cytoplasmic constituents (Huffman *et al.,* 1962; Huffman and Olive, 1964; Sinha and Ashworth, 1969). Larger than normal plurinucleate cells (with up to ten nuclei), some of which apparently result from cell fusions, have also been observed. It is also likely that, as in protostelids, plurinucleate cells develop by means of mitosis without subsequent cytokinesis. Later, the plurinucleate cells may be reduced to uninucleate ones by plasmotomy. On one occasion (Huffman *et al.,* 1962), the fusion of two cells followed immediately by nuclear fusion has been observed. The rarity of such observations leads one to doubt that this is part of a true sexual process. It is clear, however, that heteroplasmons, heterokaryons, and parasexuality may occur in the life cycle (see Section VIII). Kirk *et al.* (1971), using electron microscopy, have confirmed the presence of narrow cytoplasmic connections about 0.5 μ in width between adjoining cells in young aggregation centers, and they suggest that such connections may facilitate intercellular communication during differentiation.

One of the most intriguing aspects of cellular slime mold biology has been the question of what motivates the change in behavior that causes trophic cells to stop feeding and dividing and become attracted to each other. Earlier workers (E. W. Olive, 1902; Potts, 1902) suspected that aggregation was under chemotactic control. Evidence of this was later obtained by the demonstration that amoebae placed on both sides of a differentially permeable membrane produced aggregation streams that duplicated each other, the substance responsible for the coordinated orientation having passed through the membrane (Runyon, 1942). Bonner (1947) firmly established the existence of the chemotactic agent, which he called "acrasin." This was followed by Shaffer's demonstration (1956) that cell-free droplets from around migrating slugs, which also produced the attractant, were capable of reorienting sensitive amoebae into streams that moved toward the source of the acrasin. In a series of investigations, Shaffer (e.g., 1956, 1962) found that acrasin induced the change in surface properties of sensitive amoebae that caused them to adhere to each other, thus facilitating stream formation and morphogenesis. He concluded that cells in the aggregation centers and streams both secreted acrasin equally.

If acrasin production by amoebae is equal throughout the center and its converging streams, it is difficult without additional information to explain why the amoebae are attracted in only one direction. If one examines the excellent time-lapse motion pictures of aggregation taken by Bonner and others, it quickly becomes apparent that the cells move into the centers in rhythmic waves. Shaffer and Bonner have associated each wave with a pulse of acrasin production that begins at the center and spreads peripherally through the streams. Why, then, does the concentration of acrasin in the background not rise to a point at which amoebae become insensitive to additional pulses? In testing the acrasin-containing droplets, Shaffer found that the attractant in droplets removed from around slugs generally remains active for only a few minutes, whereas methanol extracts of acrasin are quite stable, even being resistant to boiling and exposure to acids and alkalis. He concluded that the original instability is caused by another cell product, probably an enzyme, a conclusion that has since been proved correct. The enzyme is responsible for reducing the background concentration of acrasin, thus permitting the maintenance of acrasin gradients.

Once the cells have entered the streams they become polarized and, as they move toward the center, adhere to each other anteriorly and posteriorly, probably exerting a pulling effect on each other. Shaffer refers to this as "contact following."

More recently, an exciting discovery by Konijn et al. (1967, 1969) has demonstrated that acrasin in *D. discoideum* is actually the nucleotide cyclic 3',5'-adenosine monophosphate, commonly known as cyclic AMP. This substance, discovered by Sutherland and Rall in 1957 in mammalian tissues, has been found to function in these as a "second messenger" in hormonal reactions, such as those leading to the release of free glucose from glycogen or to the synthesis of steroids. It has more recently been implicated in numerous other hormonal activities and in the control of gene expression in organisms ranging from bacteria to higher animals. It appears to be involved in many aspects of cell specialization in animals. Its action in higher animals has been postulated as follows: Any number of hormones functioning as primary messengers activate adenylcyclase at the cell surface, which converts adenosine triphosphate (ATP) into cyclic AMP and releases it within the cell. The cyclic AMP then triggers some other reaction that results in the ultimate action of the hormone (Robison et al., 1971).

Konijn et al. (1967, 1969) and Bonner (1969) have proposed that cyclic AMP during aggregation in dictyostelids, instead of being released within the cell to function as a second messenger, is released outside the cell into the medium, where it functions as a "primary messenger." The exact mechanism whereby it stimulates aggregation remains obscure. It is barely, if at all, detectable during the trophic stage and is in greatest abundance during aggregation. The nucleotide is also found in the migrating pseudoplasmodium and, during sorogenesis, in the anterior region of the sorogen, where it plays a role in the differentiation of the stalk cells.

Konijn and co-workers found that cyclic AMP is able to attract sensitive amoebae of *D. discoideum* in amounts as small as 0.01 ng in a droplet of 0.1 μl. In these experiments it was also learned that the enzyme that Shaffer thought responsible for the breakdown of the attractant in *D. discoideum* is a cyclic AMP-specific phosphodiesterase similar to but not identical with that responsible for the breakdown of cyclic AMP in mammalian cells.

Although *P. pallidum, P. violaceum,* and *D. minutum* produce extracellular cyclic AMP and intracellular phosphodiesterase, their amoebae are not attracted to the nucleotide (Konijn *et al.*, 1969; Konijn, 1972a; Bonner *et al.*, 1972). Only amoebae of *D. discoideum, D. mucoroides, D. purpureum,* and *D. rosarium* are known to aggregate toward it (Konijn, 1972a). Francis (1965) has demonstrated that *P. pallidum* produces an attractant that is not cyclic AMP, and Bonner *et al.* (1972) suggest that cellular slime molds that show no aggregation response to cyclic AMP may have different acrasins. However, cyclic AMP is thought to be active in differentiation during sorogenesis in all species, especially in stalk development. Analogs of cyclic AMP show varying degrees of effectiveness as attractants, although none is as effective as the unmodified nucleotide. The phosphate moiety of the molecule appears to be the most sensitive part (Konijn, 1972b).

A study of many isolates of dictyostelid cellular slime molds by Traub and Hohl (unpublished data) has led them to conclude that a number of *Dictyostelium* species such as *D. minutum* and *D. polycephalum* are closely related to *Polysphondylium,* in that both groups are characterized by polar spore vacuoles and an aggregation attractant that is not cyclic AMP. On the other hand, species such as *D. discoideum* and *D. mucoroides* lack polar spore vacuoles and employ cyclic AMP as the chemotactic agent. The patterns of aggregation in the two groups are also basically different. These discoveries are confirmed in part by those of Dutta and Mandel (1972), who found that the guanine–cytosine (GC) content of *Polysphondylium* is significantly greater than that of *Dictyostelium discoideum* and *D. mucoroides*.

In liquid culture, *D. discoideum, D. mucoroides, D. purpureum,* and *P. violaceum* produce extracellular phosphodiesterase (Bonner *et al.,* 1972; Gerisch *et al.,* 1972). In *D. discoideum* there is an increase in this enzyme during the growth phase until cells enter the stationary phase, at which time a heat-stable macromolecular inhibitor is released that drastically reduces the enzyme's activity (Gerisch *et al.,* 1972). Malchow *et al.* (1972) have found that amoebae of *D. discoideum* in liquid culture produce soluble extracellular phosphodiesterase primarily during the growth phase, whereas cell membrane-bound phosphodiesterase (probably the same in structure) appears at a low level by the end of the growth phase and peaks at aggregation, paralleling increased chemotactic sensitivity of the cells to cyclic AMP. They found phosphodiesterase inhibitor, which inactivates the soluble enzyme, to be relatively ineffective toward the surface-bound enzyme. They suggest that the latter may function in the chemotactic receptor system in maintaining cell sensitivity to a cyclic AMP gradient by preventing saturation of cell receptors by cyclic AMP molecules. This theory is compatible with Shaffer's earlier concept of the role of an enzyme in the chemotactic system.

Production of phosphodiesterase was found to be defective in about half of a group of morphogenetic mutants (showing fruiting aberrancies) of *D. discoideum* investigated by Riedel *et al.* (1973). Some of the mutants failed to release the inhibitor or showed delayed release; some showed excessive phosphodiesterase production; and still others were characterized by deficient enzyme activity.

Robertson *et al.* (1972b) devised an experiment in which they employed a micropipette that emitted pulses of negatively charged ions of cyclic AMP and fluorescin at intervals of about $4\frac{1}{2}$ minutes into an aqueous film containing amoebae on an agar surface. Amoebae began to be attracted to the micropipette after 4 hours and after about 6 hours were able to propagate signals themselves, with some evidence of stream formation beginning. Well-defined streams appeared after 8 hours, followed by an aggregate at the tip of the pipette. A slug developed and migrated away, later forming a sorocarp. Aggregation and pseudoplasmodial development may also occur in the presence of a constant supply of cyclic AMP.

Bonner and Dodd (1962b) and Bonner (1967) report that in a number of different dictyostelid species a more or less constant number of aggregation centers form over a given area of substrate and over a wide range of cell densities. They consider the distribution pattern to be mediated by a chemical spacing substance or "center inhibitor" acting at the substrate level. There is some evidence of more than one spacing

substance, one of which may be cyclic AMP (Kahn, 1968). After studying the effects of light on aggregation in *D. purpureum*, Kahn (1964b)
noted that light causes a reduction in territory size, perhaps by limiting
production of, or destroying, a substance that inhibits center formation
or by making the cells less sensitive to the substance.

The ability of several different species of *Dictyostelium* (*D. discoideum*, *D. mucoroides*, *D. purpureum*) to coaggregate (e.g., Bonner and
Adams, 1958) probably indicates close structural relationships in their
cyclic AMP–phosphodiesterase systems and surface properties of their
sensitized cells (Gregg, 1956; Sussman, 1966; Gerisch *et al.*, 1972). Raper
and Thom (1941) have shown that temporary cross-graftings of slugs
can be made between *D. discoideum* and *D. purpureum* and that, if
pseudoplasmodia of the two are macerated and intermixed in the absence
of bacteria, it is possible to obtain some mixed sorocarps containing
cells of both parental types. Some of the sori even contain spores of
both species. Such sorocarps are, of course, chimeric and not true hybrids.

C. Migration

At the end of aggregation in *D. discoideum* an upright cone-shaped
mass of cells develops, secretes a slime sheath around itself, and then
usually assumes a horizontal position on the substrate. This is the pseudoplasmodium or slug, which now begins migrating across the substrate
(Figs. 84A and 100). From this time on, the process of morphogenesis
proceeds as though the muticellular mass were a single individual rather
than so many separate cells. The migration time may be reduced by
increasing the ionic content (e.g., phosphate buffer) of the medium
(Slifkin and Bonner, 1952). According to Bonner and Frascella (1952),
cytokinesis, which ceases during preaggregation, begins again immediately after aggregation at a rate that rises to 1/100 cells and then falls
off to 1/200 cells during migration.

As the slug moves along, it leaves behind on the substrate a collapsed
slime sheath (Fig. 100), which is probably secreted by all the cells
(Bonner, 1967). Francis (1962) has reported that the sheath is composed of a protein–polysaccharide complex. Gerisch *et al.* (1969), from
their chemical analysis, concluded that cellulose is lacking. Recent ultrastructural studies combined with enzymatic treatments have led Hohl
and Jehli (1973) to conclude that the sheath on the substrate (but
not around the sorocarp) is an amorphous protein-containing matrix
with embedded cellulose microfibrils. In addition to its protective function, it is thought that the sheath exerts some control over migration
and morphogenesis. Because sheath material is added along the entire
length of the slug, it is thinnest at the anterior end and progressively

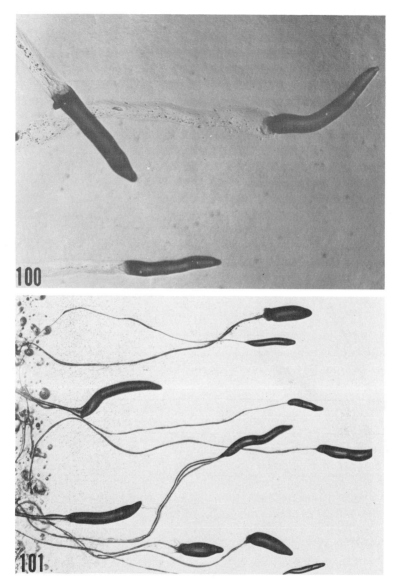

Figs. 100 and 101.   Migrating stages in *Dictyostelium*. Fig. 100. Migrating pseudo-plasmodia of *D. discoideum*. Note slime tracks, ×20. Courtesy of K. B. Raper. Fig. 101. Sorogens of *D. purpureum* migrating toward a light source and forming stalks on the substrate, ×16. From Raper (1951).

thicker posteriorly. It has been known for some time that the tip of the slug is more sensitive to certain outside stimuli, such as light and temperature gradients (Raper, 1940), and Loomis (1972) suggests that the gradient in sheath thickness is important in restricting control of migration to the pseudoplasmodial tip. He found that if the tip is removed a new one is formed anteriorly, provided the collapsed sheath does not cover the cut end. If the end does become covered, the slug reportedly can develop a new tip at the posterior end and reverse its direction of migration. (However, the possibility that the cell mass can reverse itself within the sheath is not eliminated.)

It has been demonstrated that migration stages of several dictyostelids, including *D. discoideum,* are highly sensitive to light and that quite low intensities can orient them toward the light source (Raper, 1940, 1941a). They are also attracted toward phosphorescent bacteria and luminescent paint. Light sensitivity may be related to the presence of a flavin or carotenoid (Francis, 1964). Poff and Loomis (1973) have concluded from their experiments that light increases the rate of migration by stimulating sheath formation. Strong light sources induce a reverse reaction in the slugs of *D. discoideum.* A microbeam of light focused on one side of the pseudoplasmodial tip causes the pseudoplasmodium to turn in the opposite direction and migrate away from the light source (Francis, 1964). Single amoebae do not become oriented by light. The migrating sorogens of *D. mucoroides, D. purpureum* (Fig. 101), and *P. violaceum* are light sensitive, whereas the migrating pseudoplasmodia of *D. polycephalum* are not.

The pseudoplasmodia of *D. discoideum* are also highly sensitive to heat gradients (Raper, 1940; Bonner *et al.,* 1950). Pseudoplasmodia migrate toward a warmer region over a gradient as low as 0.05°C/cm. These responses to heat gradients and light are probably of considerable survival value in enabling a species to produce its fruiting bodies in more suitable positions for spore dispersal.

A number of techniques, including the use of vital dyes such as cresyl violet and the periodic acid-Schiff reaction for polysaccharides have delimited two major regions within the migrating slug of *D. discoideum,* and the differences become more distinct as sorogenesis approaches (Bonner, 1967). The acid-Schiff reaction reveals that the cells in the posterior region, which are destined to develop into spores, contain numerous fine granules of nonstarch polysaccharide, whereas the anterior presumptive stalk cells lack them. Fluorescent spore antisera also readily distinguish between prestalk and prespore cells, the latter showing an intense fluorescence not found in prestalk cells (Takeuchi, 1963; Gregg, 1965). The fluorescence is associated with an abundance of acid mucopolysaccharide in the prespore cells (Maeda, 1971a). Also, cells of dissociated pseudoplasmodia can be separated by density centrifugation into a lighter band of prestalk cells and a heavier band of prespore cells (Maeda *et al.*, 1973). These differences continue during sorogenesis. Prespore cells are characterized by high levels of cytochrome oxidase and succinic dehydrogenase activity. A difference may also be observed in the adhesive qualities of the cells, those in the prestalk area showing a much higher degree of adhesiveness than those in the prespore region (Yabuno, 1971).

Raper (1940) showed that if a pseudoplasmodium of *D. discoideum* is transversely sectioned into several segments, all parts readjust after several hours into pseudoplasmodia with normal proportions of prestalk and prespore cells that fruit normally. If fruiting is induced to occur quickly, the anterior (prestalk) segment produces an abnormal sorocarp comprised mostly of a thick stalk and few or no spores, while the other segments fruit normally. Immunological studies on isolated prestalk segments have shown that, in the process of readjustment, those prestalk cells that later become prespore cells must first dedifferentiate into unspecialized amoebae (Gregg, 1965). Prespore cells that are disaggregated into a sparsely populated suspension dedifferentiate in about 5 hours (Gregg, 1971; Takeuchi and Sakai, 1971), and disaggregated prespore or prestalk cells in the presence of bacteria are able to dedifferentiate to the trophic stage and undergo cell division within about 9–10 hours (Takeuchi and Sakai, 1971). In dedifferentiating prespore cells the prespore vacuoles, which are unique to these cells, degenerate (Sakai and Takeuchi, 1971). The conversion of prestalk to prespore cells requires a longer period than the reverse type of conversion, since a lag phase is required for the former but not the latter (Sakai, 1973). Both types of cell conversion are blocked by inhibitors of protein or RNA synthesis.

The mechanism of pseudoplasmodial movement is still not well understood. As noted above, the slime sheath is stationary and traction for movement is between amoebae and sheath rather than with the substrate surface. Robertson and Cohen (1972) believe that the periodicity of movement observed among aggregating amoebae is maintained in the slug, which moves in successive waves of activity propagated from its anterior end and passing to the posterior end. Robertson *et al.* (1972a) emphasize that the tip of the slug from the time of late aggregation behaves as a classical organizer and is perhaps the one best understood in developmental biology. Since the tip of the slug also appears to be the prime source of its cyclic AMP, it is likely that this substance plays an important role in the slug's behavior and internal organization.

Although several factors influence duration of the migration period, none appears more important than relative humidity. When this falls below a certain point, migration ceases and the pseudoplasmodium develops into a sorocarp.

D. Culmination and Sorocarp Maturation

Sorogenesis in *D. discoideum* (Figs. 77, 84A, 102, and 103) begins when the pseudoplasmodium stops migrating and assumes an erect position on the substrate. The process of sorogenesis has been dramatically illustrated in time-interval photographs and films by Bonner and others. The erect pseudoplasmodium becomes somewhat shortened vertically just before culmination. A group of rearguard cells at the base of the pseudoplasmodium, which have the staining properties of stalk cells, remain on the substrate to form the supporting basal disc of the sorophore. As do typical stalk cells, these become highly vacuolate and secrete cellulose walls. The remaining cell mass begins to form a multicellular stalk on which the sorogen continues to move in a distal direction from the substrate until stalk formation ceases. Sorogenesis appears un-

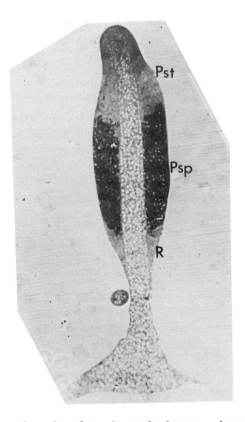

Fig. 102. *Dictyostelium discoideum;* longitudinal section of sorogen stained with Toluidine Blue. Note darkly stained prespore area (Psp), lighter stained prestalk area (Pst), and rearguard cells (R), ×130. From George *et al.* (1972).

Fig. 103. *Dictyostelium discoideum;* scanning electron micrograph of developing fruiting bodies. Note also the slime trails on substrate, ×130. Courtesy of R. P. George.

responsive to gravity and the sorocarps develop normally on upright or inverted culture plates.

Both light and electron microscopy, as well as micrurgical techniques, have contributed greatly to our knowledge of sorogenesis (Raper, 1940; Bonner, 1944, 1952; Bonner *et al.*, 1955; Raper and Fennell, 1952; Gezelius, 1959). The relationship of prespore to prestalk cells in the sorogen (as well as the slug) can readily be demonstrated by grafting experiments first designed by Raper and later used by others. Cultures of *D. discoideum* grown on the red bacterium *Serratia marcescens* have red amoebae and slugs because of the inability of the amoebae to digest the red pigment (prodigiosin). If the anterior end of such a slug is cut off and exchanged with the tip of a colorless slug grown on *Escherichia coli,* the slug with the red tip produces a red stalk and white sorus and the one with the colorless tip produces a white stalk and red sorus. The experiment clearly demonstrates that the anterior part of the slug (and sorogen) is the stalk-forming part and the posterior end is the spore-forming region. During culmination of the red-tipped pseudoplasmodium the stalk-forming red cells can be seen passing down through the center of the sorogen, the contrast being most conspicuous where they pass through the lower, nonpigmented prespore area. Regardless of the size of the pseudoplasmodium, the ratio of prestalk to prespore cells remains about 1:3 to 1:4.

Bonner (1967, 1971) has reviewed the findings of his group and others on differentiation and sorting out of cells leading up to sorogenesis. With the aid of fluorescent spore antisera, Takeuchi (1963) discovered that in *D. mucoroides* differentiation into prestalk and prespore cells begins as early as preaggregation, and at the conclusion of aggregation cells containing the spore proteins are situated at the posterior end of the cell mass and the remainder (prestalk cells) at the anterior end. According to Bonner (1967), a major rearrangement of cells occurs between aggregation and migration, and during migration of the slug the line of demarcation between prespore and prestalk regions becomes quite distinct, as demonstrated by the periodic acid-Schiff reaction for polysaccharides (Bonner *et al.*, 1955) and by other techniques.

In a strain (Ax-2) of *D. discoideum* that can be grown axenically, Garrod and Ashworth (1973) found that cells grown in the presence of glucose produce sporocarps with a stalk:spore ratio of about 1:4, but when glucose is lacking in the medium the ratio is 1:2.7. Also, when the two types of cells are mixed during the preaggregation stage, sorocarps are produced in which glucose-grown cells generally become spores and the others stalk cells. More surprising was the discovery that when tips of migrating slugs from the glucose-containing medium were used to replace tips of non-glucose-derived slugs, spores of the resultant sorocarps were derived from the glucose-grown cells. This demonstrates a developmental pattern resulting from sorting out with respect to prior growth conditions rather than the usual position effect.

Robertson and Cohen (1972), with the aid of time-lapse cinematography, have demonstrated that the pulsations that are first observed in preaggregation amoebae and that continue throughout aggregation and slug migration persist during sorogenesis. In the latter case, they are detectable as periodic jerks in the culminating sorogen and are thought to be the result of the periodic movement of prestalk cells into the developing stalk. The sorogen has a papilla-like apex and a more expanded lower region (Figs. 77C, 102, and 103). The prestalk cells in the papilla are stacked up in parallel arrangement with their long axes at right angles to the stalk, resembling in section cells in the sorogen of *Acytostelium*. These cells, the inner portions of which now contain an abundance of nonstarch polysaccharide, secrete a stalk tube of cellulose before passing down into the tube at the apical end of the papilla and joining other cells that have preceded them in stalk development (Fig. 77, B and C). As they do so, they become distinctly vacuolate. The first stalk cells that pass through the center of the sorogen eventually reach the basal disc, thus completing the union between stalk and disc. It is at this point that the sorogen begins to rise from the substrate and the stalk becomes externally visible. Some rearguard cells that fail to remain with the basal disc are carried up at the base of the sorogen (George et al., 1972). Such stains as Toluidine Blue clearly delimit prestalk and prespore regions of the sorogen (Fig. 102). These developments are discussed more fully by Bonner (1967), who made many of the original investigations.

As the developing stalk cells pass below the papilla region, they secrete an additional layer of cellulose on the inside of the original tube and then they secrete their own cell walls (Gezelius, 1959). The stalk cells become nonviable even within the area of the sorogen, where the nuclei begin to degenerate. The mature cells appear empty and expanded, with their cellulose walls polygonally flattened against each other to the exclusion of intercellular spaces. The stalk tapers from base to apex, and smaller fruiting bodies may have only a single row of cylindrical cells above.

As the last of the prestalk cells are used up the papilla disappears. The prespore cells, which have already begun their transformation into spores, complete their encystment. The spore case is composed primarily of cellulose. The opaque sorus, which is white to faintly yellow in D. discoideum, consists of short-cylindrical to bean-shape spores in a mucilaginous matrix. During sorogenesis the developing sorocarp is surrounded by a thin slime sheath, which breaks down around the sorus as the spores mature.

There is now evidence that cyclic AMP has a controlling role in the differentiation of stalk cells in Dictyostelium, perhaps here as a secondary messenger. As previously noted, cyclic AMP production in the slug is concentrated primarily at its tip (Bonner, 1949). Recently, Bonner (1970) discovered that amoebae of D. discoideum, in the presence of 10^{-3} M cyclic AMP, develop singly or in groups into stalk cells within 48 hours on an agar substrate. Since it is known that this substance is produced in the sorogen only in the region of stalk formation, it was concluded that it is a key factor in stalk cell differentiation. On the other hand, cyclic AMP in excessive concentrations interferes with normal movement and polarity relationships within the sorogen and thus disrupts normal fruiting (Nestle and Sussman, 1972). If added early enough in sorogenesis it completely or partially inhibits stalk development, probably by interfering with the normal gradient of cyclic AMP in the sorogen. Two enzymes known to be important in morphogenesis are also drastically inhibited. Detailed reviews of the biochemistry of differentiation and morphogenesis in Dictyostelium have been published by Ashworth (1971), Newell (1971), and Wright (1973).

A spacing substance, different from that affecting aggregation centers, has been implicated in the development of dictyostelid fruiting bodies (Bonner and Dodd, 1962b; Bonner, 1967). Adjacent sorocarps repel one another during culmination, their stalks bending away from each other. If a glass rod is inserted in the agar near a developing sorocarp, the stalk assumes a position bisecting the angle between substrate and rod. A suspicion that the spacing substance is a gas has been supported by the discovery that charcoal placed on the substrate to one side of a developing sorocarp causes it to curve over into it, indicating that a gas causing the spacing is being absorbed by the charcoal. The nature of the gas has not been determined.

E. Macrocyst Formation

Under certain conditions unfavorable to sorogenesis, cell aggregates of some dictyostelid isolates are able to enter into an alternative developmental pathway resulting in the formation of macrocysts. Macrocysts (Fig. 87), first discovered in *Dictyostelium mucoroides* (Raper, 1951; Blaskovics and Raper, 1957), have now been found in *D. discoideum, D. purpureum, D. minutum,* and *Polysphondylium violaceum.* Macrocyst formation is favored by growth in the dark, under moist conditions (Weinkauff and Filosa, 1965) and on a medium lacking phosphates (Nickerson and Raper, 1973). Under these conditions the aggregates, or subdivisions of them, become surrounded by thick walls and enter a resting stage. On germinating, they produce numerous amoebae. Their ontogeny has been elucidated by ultrastructural studies (Section VII).

VII. ULTRASTRUCTURE

The trophic stage of dictyostelids reveals, for the most part, an ultrastructure typical of eukaryotic cells (see Mercer and Shaffer, 1960; Hohl *et al.,* 1968). The mitochondria have tubular cristae. Mitosis is intranuclear (Fig. 107). Investigations of changes that occur during morphogenesis have been of particular interest. Although *D. discoideum* has again been the favored research subject, some attention has also been given to *Acytostelium* and *Polysphondylium.*

A. Acytostelium

The ultrastructure of *A. leptosomum* was investigated by Gezelius (1959) and Hohl *et al.* (1968), who found that the stalk tube contains parallel cellulose fibrils longitudinally oriented. In her comparative study of stalk development in *Acytostelium* and *Dictyostelium,* Gezelius concluded that the stalk tube of the former is homologous with the outer layer of the stalk tube of the latter. In median vertical sections of the sorogen of *A. leptosomum* the cells in section (except for the cap cell) all appear elongate, with their long axes at right angles to the stalk (Fig. 83). There is a microfibrillar, pseudopod-like zone at each end of the cell, one abutting on the stalk and the' other on the slime sheath in a manner indicating their involvement in the upward movement of the sorogen (Fig. 104). Bundles of microfibrils traverse the cells longitudinally and are also probably involved in cell movement. Microtubules

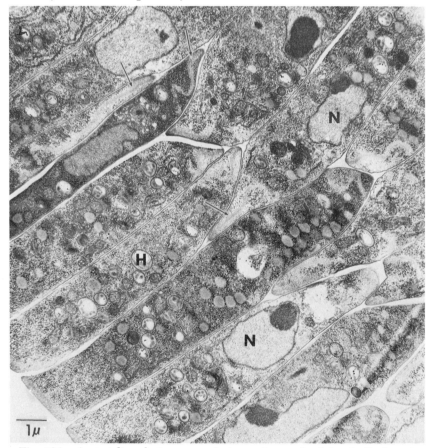

Fig. 104. *Acytostelium leptosomum;* electron micrograph of longitudinal section through sorogen, showing pseudopodium-like cell extremities, fibrillar strands (arrows), nuclei (N), and "H bodies" (H). Stalk not included in this section. From Hohl *et al.* (1968).

are present but randomly oriented. A variety of vesicles, multivesicular bodies, lysosome-like bodies, and "H bodies," which appear to have the characteristics of prespore vesicles later found in *Dictyostelium,* are arranged across each cell and are thought to be involved in producing and transferring structural material to the stalk. Hohl *et al.* have concluded that the fibrils for stalk formation are assembled just inside the plasmalemma, which they penetrate before streaming away to become part of the stalk tube. Before the cells mature into spores, a number of cytoplasmic organelles are taken into autophagic vacuoles and digested. All cells of the sorogen develop into globose spores, which have a heavy outer cellulose layer and two thin inner bands reported to be formed on the cytoplasmic side of the plasmalemma. Microcysts, which are also round, have similar walls.

B. *Dictyostelium* and *Polysphondylium*

1. The Trophic Stage

The studies by Peter Roos (unpublished data) of the University of Geneva on *Polysphondylium violaceum* have produced the most detailed information on ultrastructure of the dictyostelid trophic stage. Engulfment of food bacteria is accomplished by means of pseudopodial processes (Fig. 105). During mitosis an intranuclear spindle develops with numerous well-defined microtubules extending between two polar electron-dense microtubule-organizing centers (MTOC's) just outside the nuclear membrane (Fig. 107). The spindle microtubules penetrate the nuclear membrane at the poles. Astral microtubules radiate from the MTOC's into the cytoplasm. The MTOC's have no discernable internal structure. On rare occasions a structure is found that in some ways resembles a vestigial kinetosome associated with cytoplasmic microtubules and a beaked portion of the nucleus (Fig. 106). Further information on the nature of this interesting body is needed.

2. Sorogenesis

The most extensive ultrastructural studies of morphogenesis are those by George (1968), Hohl and Hamamoto (1969a), Ashworth *et al.* (1969), and George *et al.* (1972) on *Dictyostelium discoideum.* The discovery of ultrastructurally detectable alterations in the plasmalemma of amoebae induced by cyclic AMP and calcium (Gregg and Nesom, 1973; Aldrich and Gregg, 1973) may help to explain previously discussed changes in the behavior of amoebae approaching aggregation and is

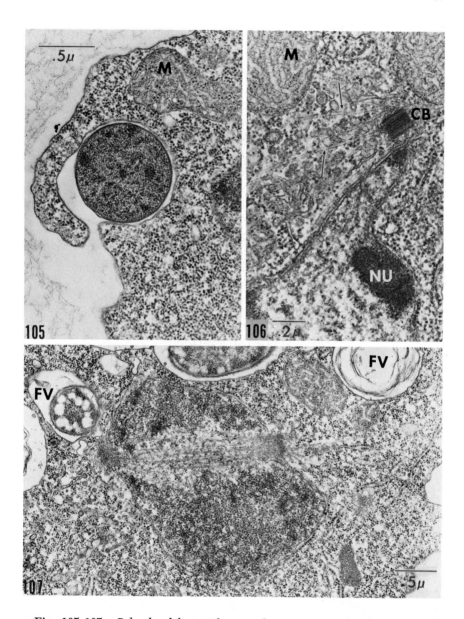

Figs. 105–107. *Polysphondylium violaceum;* electron micrographs of amoebae. Fig. 105. Engulfment of bacterial cell. Fig. 106. Beaked end of nucleus with nearby centriole-like body associated with microtubules. Fig. 107. Metaphase of mitosis, showing intranuclear spindle, dense polar areas with astral microtubules, and intact nuclear envelope. M, mitochondria; CB, centriole-like body; NU, peripheral nucleolus; FV, food vacuoles. Courtesy of U.-P. Roos.

in agreement with evidence from the use of labeled antibody that a special class of membrane sites are identifiable on aggregating cells (Beug *et al.*, 1973a,b). Cell aggregration is accompanied by the appearance of autophagic vacuoles replacing food vacuoles, and autodigestion is now probably a major source of energy in these cells (Maeda and Takeuchi, 1969).

Special vacuoles known as "prespore vacuoles" (Fig. 108), which appear in cells of the prespore region of the migrating slug, were discovered independently by Hohl and Hamamoto (1969a) and Maeda and Takeuchi (1969). The prespore vacuole is bounded by a unit membrane lined internally by a dense band of material and enclosing an aggregate of fine filaments. The evidence strongly indicates that they are the sites of the nonstarch polysaccharide mentioned in Section VI. The membrane of the vacuole fuses with the plasmalemma and the expelled vacuolar contents contribute to formation of the spore wall and surrounding mucilaginous matrix (Takeuchi, 1972). This interpretation is consistent with the information from immunohistochemical tests (Takeuchi, 1963; Ikeda and Takeuchi, 1971; Maeda, 1971a).

It has been proposed that prespore vacuoles originate from Golgi

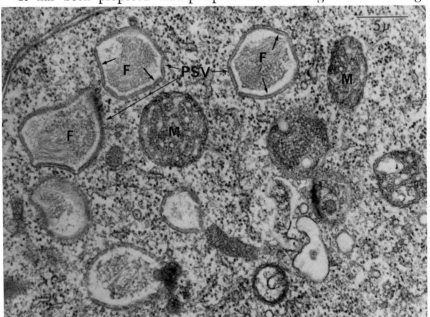

Fig. 108. *Dictyostelium discoideum;* electron micrograph of prespore cell showing prespore vacuoles (PSV), the unit membrane of which is lined internally with thin electron-dense layer (arrows) enclosing a filamentous aggregate (F). Mitochondria (M) have reticulate, tubular cristae. From Maeda (1971b).

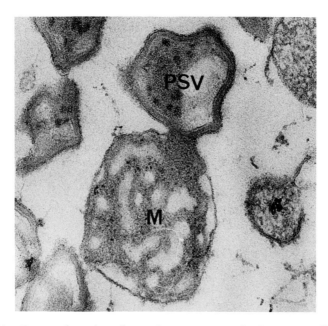

Fig. 109. *Dictyostelium discoideum;* electron micrograph of prespore cell showing prespore vacuole (PSV) apparently developing from padt of a mitochondrion (M). From Maeda (1971b).

bodies (Hohl and Hamamoto, 1969a) and, alternatively, that they are derived from evaginations in the outer membranes of mitochondria or from part of the main body of a mitochondrion (Fig. 109) and contain the mitochondrial enzyme succinic dehydrogenase (Maeda, 1971b). A second type of vacuole, called the "spore vacuole," has been reported in the prespore cells and spores during sorogenesis (Gregg and Badman, 1970; Gregg, 1971). These vacuoles are thought to contain the trehalose detected earlier in mature spores (Clegg and Filosa, 1961). Neither type of vacuole is found in prestalk cells.

Electron microscopy has clearly shown that prestalk cells are in close contact with each other and lack intercellular spaces, whereas prespore cells are more loosely packed and have intercellular spaces. This agrees with the demonstration that prestalk cells exhibit a much higher degree of adhesiveness than prespore cells (Yabuno, 1971).

The sheath around the sorocarp is an amorphous protein-containing matrix which, unlike that produced on the substrate during migration, lacks cellulose microfibrils (Hohl and Jehli, 1973). Sections through the papilla of the sorogen show that the prestalk cells are elongate and horizontally oriented (Fig. 110), appearing much like cells in sec-

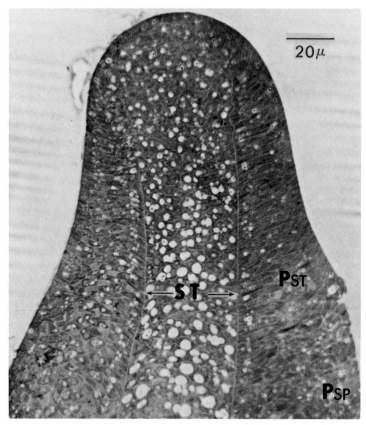

Fig. 110. *Dictyostelium discoideum;* light micrograph of papilla of sorogen, showing prespore cells (Psp); transversely elongated prestalk cells (Pst), those nearest
stalk showing parallel alignment of vacuoles; and stalk tube (ST) enclosing highly
vacuolate stalk cells. From George *et al.* (1972).

tions through *Acytostelium* sorogens. Their contents are more or less
stratified from stalk to slime sheath in the following order: fibrillar pseudopodial zone, mitochondrial zone, vacuolar zone with prominent autophagic vacuoles, nucleus, and unstratified cytoplasm. The pseudopodial
zones are oriented against the stalk in a manner indicating that they
contribute to the upward movement of the sorogen (Fig. 111). This
observation, in addition to the known increase in adhesiveness of these
cells in relation to each other and to nonliving surfaces (in this case
the stalk and probably the slime sheath), may go far toward explaining
how the sorogen moves up the stalk.

The inner layer of the stalk tube is not present in the papilla but begins to form further down, being deposited on the inside of the outer tube layer after the cells have ceased moving and are about to form their own polygonal cell walls (Gezelius, 1959; George *et al.*, 1972) (Fig. 77C). Because the cells are not moving when the inner layer is produced, the cellulose fibrils are randomly oriented, in contrast to the parallel and vertical arrangement of fibrils in the outer layer produced by cells during vertical migration. The stalk cell walls also consist of randomly oriented fibrils. The elementary fibrils of the stalk contain crystalline micelles characteristic of cellulose fibrils.

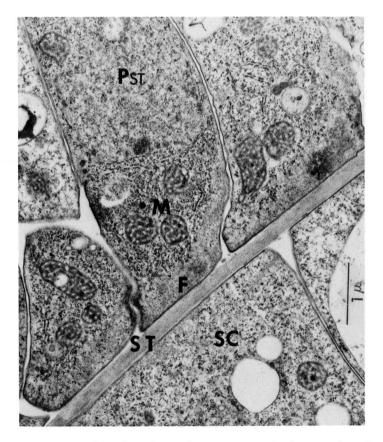

Fig. 111. *Dictyostelium discoideum;* electron micrograph of section through portion of sorogen. Elongate prestalk cells (Pst) at upper left are migrating toward upper right-hand corner along outer stalk tube layer (ST) of parallel cellulose fibrils. Note fibrous cytoplasm (F) of pseudopodium-like areas of prestalk cells, mitochondria (M), and vacuolate stalk cells (SC). From George *et al.* (1972).

A radioautographic study with labeled glucose has shown that this carbohydrate is rapidly incorporated into the stalk tube but not into the sorocarp's slime sheath, in which cellulose is generally conceded to be absent (George, 1968). Surprisingly, even though the mature stalk contains only dead and essentially empty cells, it is able to convert labeled glucose into polysaccharides days after sorocarp maturation, presumably by means of enzymes and high-energy precursors that persist in the stalk region. Calcium, which is in higher concentration in prestalk cells of the slug than in prespore cells, is known to influence cell differentiation leading to stalk formation (Maeda and Maeda, 1973).

The mature spore wall was at first thought to be three-layered, but a recent freeze-etch study, combined with enzymatic tests, has shown that there are four layers: an outer mucopolysaccharide layer, two middle cellulosic layers with their cellulose fibrils differently oriented, and an inner layer of cellulose and protein (Hemmes et al., 1972). During spore germination the outer wall layers rupture as the spore swells, following which the inner wall is breached and the amoeba emerges (Cotter et al., 1969). Crystals and lipid bodies in the cytoplasm disappear, and the mitochondria become less dense.

3. Macrocysts

Macrocysts are produced by certain isolates of D. mucoroides, D. minutum, D. purpureum, D. discoideum, and P. violaceum. They are a type of resting stage formed from aggregates of amoebae that become rounded and enveloped by heavy cellulose walls (Blaskovics and Raper, 1957). On germination they produce trophic amoebae. The developmental details of macrocyst development have only recently become available. Filosa and Dengler (1972), working with D. mucoroides, and Erdos et al. (1972), using P. violaceum, have shown that a fibrillar sheathlike layer is first laid down around each cell cluster that is to become a macrocyst. Meanwhile, a larger than normal amoeboid cell appears within each cluster and proceeds to phagocytize the other amoebae (Fig. 112). The larger amoeba is reported to be uninucleate in D. mucoroides but first binucleate and then uninucleate in P. violaceum, where it is suspected of originating by cell fusion followed by nuclear fusion.

After all or most of the smaller amoebae have been ingested, a firm secondary wall develops around the macrocyst (Erdos et al., 1972, 1973a), and this is followed by a trilaminar tertiary wall (the innermost wall of the mature macrocyst). The engulfed amoebae degenerate and the single remaining giant cell now fills the interior of the macrocyst (Fig. 113). It contains a single large, multinucleolate nucleus, in which

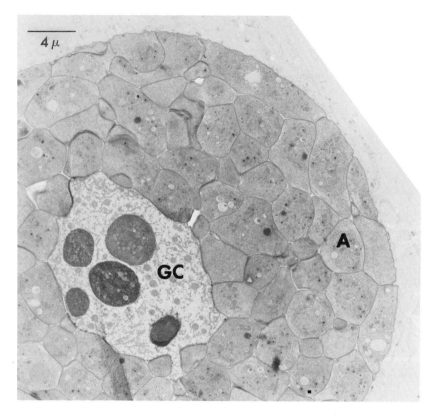

Fig. 112. *Polysphondylium violaceum;* electron micrograph of section through macrocyst, showing central giant cell (GC) with ingested amoebae, surrounded by aggregate of amoebae (A). Primary wall has begun to form. From Erdos *et al.* (1972).

Erdos *et al.* (1972) have found axial elements characteristic of synaptic chromosomes. The latter occur singly or in pairs or groups but have failed to show evidence of central elements. This interesting discovery, supported by the recovery of a sizeable group of recombinant genotypes from heterozygous macrocysts (G. W. Erdos, personal communication), indicates that the macrocysts may be the sites of a sexual process in dictyostelids. The large size of the nucleus leads one to suspect a high degree of polyploidy. The macrocyst cell becomes multinucleate before complete dormancy occurs.

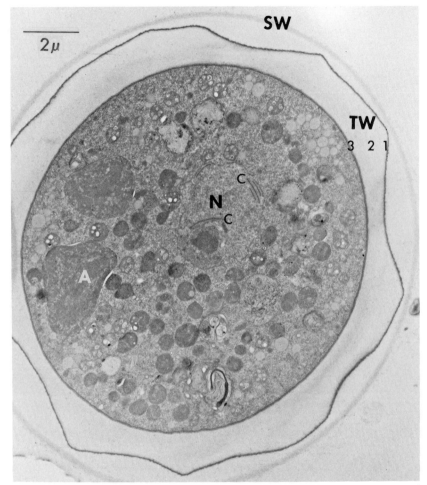

Fig. 113. *Polysphondylium violaceum;* electron micrograph of a section through a mature macrocyst. Some ingested amoebae (A) are still apparent. The large nucleus (N), thought to be in meiotic prophase, contains what appear to be synaptic chromosomes (C). The secondary wall (SW) and a three-layered tertiary wall (TW) surround the macrocyst. From Erdos *et al.* (1972).

During germination in *D. mucoroides* (Erdos *et al.*, 1973b), the large multinucleate protoplast of the macrocyst cleaves progressively into relatively large, uninucleate "proamoebae." This is followed by cell divisions of the latter and the production of numerous trophic amoebae of normal size that escape as the macrocyst walls break down.

The ultrastructure of microcyst development and germination in *Polysphondylium* has been investigated by Hohl *et al.* (1970), who found the mature wall to be two-layered.

VIII. GENETICS

While evidence of a parasexual process in dictyostelids has been accumulating for a number of years, it had been generally thought that conventional sexuality was lacking (Olive, 1963). However, the question of sexuality has been reopened by the discovery under the electron microscope of evidence of synapsis. Adding to this are the reports that recombinant genotypes were obtained from germinating macrocysts (G. W. Erdos, personal communication) and that macrocyst production in *D. discoideum, D. purpureum,* and *Polysphondylium* is controlled by what appears to be a one-locus, two-allele mating type system (Erdos *et al.,* 1973b; Clark *et al.,* 1973). However, conclusions as to whether a typical sexual process is involved must await thorough genetic analysis of progeny from germinating macrocysts and the demonstration that the so-called "mating types" do not represent isolates carrying a variety of mutant factors blocking macrocyst development. Some single-cell strains of species purported to have mating types are able to produce macrocysts when grown alone and in one such species, *P. violaceum,* some strains may cross with either mating type. Genetic interpretations are also likely to be complicated by the propensity toward aneuploidy in these organisms.

Bonner (1967) has reviewed the work on cell variation and mutation in dictyostelids and recent investigators have contributed further to this subject. The most successful mutagenic treatments have been with ultraviolet irradiation (e.g., Sussman, 1955; Sussman and Sussman, 1953) and nitrosoguanadine (Loomis and Ashworth, 1968; Yanagisawa *et al.* 1967; Fukui and Takeuchi, 1971). Most of the mutants studied have been those that fail to aggregate (*aggregateless*), or that aggregate without fruiting (*fruitless*), or that produce aberrant sorocarps (*sporeless, bushy, dwarf*). More recently, drug-resistant and temperature-sensitive mutants have been investigated.

Sussman (1961b) was probably the first to demonstrate the recovery of recombinant phenotypes in dictyostelids. Using a yellowish-spored wild-type isolate of *D. discoideum,* he obtained a white-spored mutant and a yellowish-brown one, the latter at first resembling wild type but later forming a brownish pigment, some of which diffuses into the medium. On mixing these two mutant phenotypes, he recovered from 2600 spores a single apparent wild-type heterozygote that later gave rise to parental and recombinant (yellow and white-brown) phenotypes. The low frequency of heterozygote formation has been interpreted as indicating parasexuality.

Further evidence of parasexuality has come from several independent investigations. Loomis and Ashworth (1968) studied several mutant strains of *D. discoideum*, including five nitrosoguanadine-induced morphological mutants, an actidione-resistant one, and Sussman's brown mutant, in various paired combinations and recovered wild types at a frequency of 1.3/100 to 7.3/10,000. Although reciprocal recombination was not demonstrated, the resultant phenotypes obtained by plating out the wild types indicated that some form of parasexual recombination had occurred, though some of the wild-type frequencies were higher than expected from a parasexual process.

Sinha and Ashworth (1969) studied similar mutants in five-point crosses in conjunction with chromosome counts and reported the recovery of diploids ($2n = 14$) at the rate of 1/3000. These proved to be unstable, breaking down through various levels of aneuploidy to the haploid condition. Numerous recombinant phenotypes representing ten different kinds of segregants were recovered. Evidence of chromosomal exchanges accompanying haploidization was not obtained. The authors concluded that cell anastomoses and fusions (see Section VI,B) permitted the association and occasional fusion of nuclei of different amoebae. Fukui and Takeuchi (1971) studied combinations of mutants for drug resistance and pigmentation and concluded that the heterozygotes which appeared in their liquid shake cultures developed only during the stationary culture phase, when the amoebae became relatively inactive and were more likely to form stable cell fusions that result in heterokaryons. They theorized that chromosomal mixing, instead of occurring by conventional nuclear fusion, might ensue from the division of the heterokaryon on a single large spindle, as was demonstrated in animal tissue culture cells (Fell and Hughes, 1949).

The first evidence of linkage groups in a dictyostelid has been obtained by Katz and Sussman (1972) by means of selective techniques based on the use of temperature-sensitive and cycloheximide-resistant mutants of *D. discoideum*. Heterozygotes appeared at the rate of about 10^{-5}, and segregants from the latter at about 10^{-3}. Three different linkage groups were established from six-point crosses. The actual mechanism of the diploidization and haploidization processes responsible for the genetic results remains obscure. Katz and Sussman, while agreeing that differential chromosomal losses occur, find that their diploids are much more stable than those of Sinha and Ashworth and conclude that instability of the aneuploid condition often results in reattainment of the diplophase, possibly by preferential duplication of the sister chromosome to the one lost (resulting in homozygosity for any genes on that chromosome). These and the foregoing results confirm the occurrence of some kind of parasexual process in dictyostelids. With the aid of the newly discovered linkage groups it should now be possible to determine whether exchanges between homologous chromosomes also occur.

Recently, genes for axenic growth in *D. discoideum* have been found on two of the seven chromosomes (Williams *et al.*, 1974). The usefulness of this species in biochemical and genetic studies has been greatly enhanced by the discovery that these recessive genes can be bred into other strains via the parasexual process and that both haploid and diploid axenic strains carrying newly introduced markers may thereby be obtained. These strains are now more susceptible to analysis under axenic culture conditions.

Synergism, a kind of "cross-feeding" between cells, may occur between certain paired mutants of a dictyostelid. For example, when a fruitless and an aggregateless strain of *D. discoideum* were mixed, normal sorocarps were produced (Sussman, 1954). Only the mutant phenotypes were recovered from the spores. The lack of a synergistic response when two such mutants are placed on opposite sides of a thin agar membrane indicates that the cells need to be in contact to permit an exchange of cell products through the plasma membranes or of cytoplasm through cell anastomoses. Yamada *et al.* (1973) studied 71 developmental mutants of *D. purpureum*, which they classified as (1) aggregateless, (2) aggregating but nonfruiting, and (3) aggregating and forming aberrant finger-shaped sporeless columns. Interclass but not intraclass mixtures yielded fruiting bodies with spores. The three developmental stages were thought to be controlled by three different gene groups.

Kahn (1964a) investigated an unusual aggregateless mutant of *D. purpureum* and found that its amoebae responded to the introduction of relatively few wild-type cells by aggregating and forming normal sorocarps. The spores of these sorocarps gave rise only to wild-type cultures, showing that in some way the association of aggregateless with wild-type cells had repaired the deficiency of the former. Filosa (1960) discovered that a fruitless mutant of *D. mucoroides* could be restored to normality, including normal sorocarp development, if ethionine was added during the trophic stage. Two aggregateless mutants of *D. discoideum* studied by Weber and Raper (1972) produced mature sorocarps when either was brought into contact with other types of mutants or with wild type of the same species. However, only the original aggregateless phenotypes were recovered from the spores. A similar induction of fruiting occurred when either aggregateless mutant was placed in contact with wild-type amoebae of *D. mucoroides* or *D. purpureum*, but normal fruiting failed to occur with *Polysphondylium pallidum*. This is consistent with the known ability of these three species of *Dictyoste-*

lium to coaggregate with each other but not with *P. pallidum,* indicating a closer relationship among the three species of *Dictyostelium* than any of them shows with *P. pallidum.*

There has been a recent interest in cellular slime molds as tools for the investigation of gene activity at both the translational and transcriptional levels during morphogenesis (Firtel, 1972; Firtel and Bonner, 1972; Firtel *et al.,* 1972). Analysis of the DNA of *D. discoideum* by DNA–DNA renaturation kinetics has shown that about 30% of the nucleotide sequences are repetitive. DNA–RNA hybridization experiments strongly indicate that single-copy rather than repetitive DNA is more active during the various morphogenetic stages. At successive stages of morphogenesis both qualitative and quantitative changes in the portion of single-copy DNA being transcribed can be detected, the specific activities of certain enzymes gradually increasing at a particular stage and then decreasing. It is estimated that the DNA transcript is large enough to code for over 8000 proteins the size of globin chains, indicating an enormous diversity of sequences required to regulate differentiation in this relatively simple eukaryote. Even so, the complexity of the DNA is considered low in comparison with a eukaryote, such as *Drosophila,* in which the genome is two to three times greater.

Certain aspects of these studies can probably be facilitated by the use of such forms as protostelids, *Acytostelium,* and certain acrasid cellular slime molds (e.g., *Copromyxa*) which appear to have only one type of cell present during morphogenesis rather than a combination of prestalk and prespore cells.

CHAPTER 4

Myxogastria (Myxomycetes)

I. INTRODUCTION

The Myxogastria are more commonly known as the myxomycetes or true slime molds. Haeckel (1868) placed the group in the Protista. De Bary (1859, 1887) was probably the first to understand the life cycle of a myxomycete. He also recognized the animal-like characteristics of the group and believed that they had evolved independently of the fungi from simpler flagellate organisms. He noted the superficial nature of resemblances between mycetozoans and fungi, pointing out that a cell wall is absent in the trophic stage of slime molds but present in that of the fungi. Major differences in modes of nutrition—primarily phagotrophic in slime molds and osmotrophic in fungi—have long been recognized. In addition, the flagellar apparatus is unlike that of any group of true fungi. Nevertheless, because of their fungus-like appearance and their general neglect by zoologists, the slime molds have been studied chiefly by mycologists, many of whom have classified them among the fungi. Martin (1960) and Martin and Alexopoulos (1969) place them in a special subdivision of the fungi called Myxomycotina. Recent investigations in our laboratory indicate the origin of myxomycetes from the protostelids.

In spite of considerable diversity in the appearance of trophic and fruiting stages, the myxomycetes show a remarkable degree of homogeneity in their life cycles, which involve an alternation of haploid and diploid phases. Uninucleate, haploid flagellate cells resembling those of flagellate protostelids are present and readily convert into nonflagellate myxamoebae. In most species that have been studied there is a sexual process that leads to the production of a multinucleate diploid trophic stage known as the "plasmodium." Both haploid and diploid trophic stages phagocytize other living organisms, especially bacteria and fungi.

99

Typical myxomycete plasmodia differ from those of protostelids in showing a rhythmic, reversible flow or shuttle movement of protoplasm. They differ from the pseudoplasmodia of cellular slime molds in being coenocytic rather than cellular bodies. The plasmodia produce the fruiting bodies, in which the spores are cleaved endogenously, with meiosis occurring in the spores.

The biology of the myxomycetes has recently been covered in considerable detail by Gray and Alexopoulos (1968) and their taxonomy, by Martin and Alexopoulos (1969), to which publications the reader is referred for information beyond the scope of the present treatment. We acknowledge the important contribution of these excellent texts to the preparation of this chapter, in which we emphasize primarily the more recent concepts of myxomycete biology. For more detailed information on the background of investigations leading up to our current understanding of the group, the above-mentioned texts should be consulted in addition to the very helpful reviews of Alexopoulos (1963, 1966, 1969).

A large body of information has accumulated in recent years on the biochemistry, morphogenesis mechanism of movement, and the nature of sexual and somatic compatibility in myxomycetes. Because of the unique features of their life cycles and the ease with which a number of species may be cultured in the laboratory, the slime molds have become increasingly popular experimental subjects. The recent extensive literature on the biochemistry of their differentiation, especially at the macromolecular level, has been reviewed by Sauer (1973).

II. OCCURRENCE, ISOLATION, AND MAINTENANCE IN CULTURE

A. Occurrence and Collection

Between 400 and 500 species of myxomycetes are known at the present time. The sporocarps of most of them can be seen with the unaided eye. They commonly occur on such substrates as decaying logs, humus, soil, and the bark of living trees. Some occur on dung and others may even appear on living leaves. Such substrates as the latter do not support growth of the organism but are merely convenient substrates for fruiting. In our own experience, we have also found certain myxomycetes fairly common on moribund plant structures, such as dead inflorescences, old capsules and pods, and fleshy fruits still attached to the plant. Among the most productive of these substrates for myxomycetes and other mycetozoans are old cattail inflorescences. The smallest sporocarps are not observable without the aid of a microscope, whereas others may reach a breadth of over 1 foot. The majority are measured in millimeters.

Myxomycete sporocarps show a wide range of shapes and colors. They may appear as relatively small, stalked or sessile sporangia, most often in clusters, or as sessile irregular branched or reticulate masses, or as globoid or hemispherical mounds that are often quite large and conspicuous. Some of the characteristic colors are white, yellow, orange, red, brown, and black.

Both hand lens and pocket knife are useful for collecting purposes. Portions of the substrate bearing the sporocarps may be placed in small boxes, such as pillboxes, and glued to their bottoms for safe return to the laboratory. Plasmodial stages are often found under loose bark, beneath a log, or within the rotting wood. Those most likely to be noticed generally appear as glistening, flat, fan-shaped, and usually reticulate masses that are most often yellow or white in color but may also be orange, red, brown, or various shades of violet to black.

The collection of myxomycetes in temperate zones is best during the summer months, but even in winter plasmodia and subsequently fruiting bodies of a number of species may be obtained by plating out likely substrates onto suitable agar media or simply by incubating the substrates in moist chambers. A method that generally yields a variety of species with small sporocarps is to bring in bark collections from a variety of living trees, wet them, and place them in a moist chamber for several days, following which an examination of the substrate with a dissecting microscope often reveals the presence of small fruiting bodies (Gilbert and Martin, 1933). The same may be done with a variety of other substrates, such as dead leaves, twigs, and dung (Alexopoulos, 1964).

It is of interest to note that some myxomycetes have been found to be parasitized by fungi. Several species of the pyrenomycetous genus *Nectria* have been found on the fruiting bodies of a number of myxomycetes (Samuels, 1973).

B. Isolation and Culture

Gray and Alexopoulos (1968) have listed the species of myxomycetes that have been induced to complete their life cycles in culture, these amounting to somewhat less than 15% of all known species (see also Indira, 1969). A variety of synthetic media, in conjunction with a food organism, have been found suitable for laboratory culture. These include hay infusion agar, oat-flake agar, and cornmeal agar (see Appendix). Plasmodia may generally be isolated from the natural substrate by placing a portion of the latter on the agar surface. The plasmodium usually migrates onto the agar surface, after which it may be transferred to a fresh culture dish and supplied with sterile pulverized oats sprinkled onto the agar. A migrating plasmodium, on reaching the ground oats, tends to accumulate around this food source and to ingest it. The plasmodium is likely to carry with it certain bacteria from the natural substrate, which generally do not interfere with its development but rather contribute to the food supply.

Cultures may also be obtained by spreading spores taken from sporocarps onto the surface of the agar medium. As the spores germinate, sufficient native bacteria may develop to make it unnecessary for more to be added. Otherwise, an inoculum of *Escherichia coli, Aerobacter aerogenes*, or some other bacterium should be added. Again, pulverized oats may be supplied after the first small plasmodia have appeared. After mature plasmodia have developed, fruiting may often be induced by transferring them to nonnutrient media. Species with yellow plasmodia require light for fruiting. Further information on the induction of sporocarp development is given in Section VI,D.

In our own experience, monoxenic cultures on a known bacterium may most readily be obtained by placing a large number of spores on an agar medium suitable for their germination. As the myxamoebae appear and multiply, many of them migrate away from the inoculum and onto the sterile portion of the agar surface. These may then be isolated by cutting them out with a sharp spearpoint needle on a small block of agar and transferring them to a fresh plate, where a small amount of the desired bacterium can be added. It is advisable to leave some transfers uninoculated to check for the possibility of residual bacterial growth. Additional insurance against residual bacterial contamination may be obtained by placing the spores on an agar medium to which 2×10^6 units of phosphate penicillin G and 1 gm of streptomycin sulfate per liter have been added. These antibiotics inhibit most bacteria but not the myxamoebae. Rigorous methods for obtaining cultures of myxamoebae and plasmodia free of contaminating organisms are discussed in greater detail by Gray and Alexopoulos (1968).

A number of slime mold species have now been grown in axenic culture. Daniel and Rusch (1961) have grown plasmodia of *Physarum polycephalum* axenically on a partially defined liquid medium containing chick embryo extract, yeast extract, tryptone, and inorganic salts. Later, hematin was found to be a satisfactory substitute for chick embryo extract (Daniel *et al.*, 1962). Both the myxamoebal and plasmodial stages have been maintained in axenic culture. Goodman (1972) was able to maintain the myxamoebal stage in culture for more than 100 serial transfers. Ross (1964) grew plasmodia of *P. flavicomum* and *Physarella oblonga*, as well as myxamoebae of *Badhamia curtisii*, on similar media, although none was induced to complete its entire life cycle in axenic culture. On the other hand, *Physarum compressum*, which appeared as a contaminant, fruited under these circumstances. Schuster (1964b) cultured amoebae of *Didymium nigripes* and *Physarum cinereum* on killed bacteria, and the latter species formed nonsporulating plasmodia. Several later publications (H. R. Henney and Henney, 1968; Henney and Lynch, 1969; Haskins, 1970) have described axenic growth of additional species.

Both *P. polycephalum* (Daniel and Baldwin, 1964) and *P. flavicomum* (Ross and Sunshine, 1965), were induced to complete their life cycles in axenic culture, but the myxamoebal stage was absent. The three species of myxomycetes that have been grown on chemically defined media are *P. polycephalum* (Daniel *et al.*, 1963), *P. flavicomum*, and *P. rigidum* (Henney and Lynch, 1969). All three species require hematin, biotin, and thiamine, and several amino acids are included in the media.

It cannot be concluded that plasmodia grown axenically are characterized exclusively by osmotrophic nutrition. Guttes and Guttes (1960) have demonstrated that plasmodia in liquid culture engulf droplets of the medium by means of pinocytosis. It is probable that plasmodia in nature are nourished both by ingestion of food organisms and by pinocytosis and diffusion through the plasma membrane.

III. LIFE CYCLE OF A MYXOMYCETE

The life cycle of a myxomycete may be completed within a period of several days to a few weeks in culture, depending on the species and the conditions of culture. A typical life cycle is diagrammed in Fig. 114. The form illustrated is heterothallic. Although there has been much debate in the past about where karyogamy and meiosis occur, recent ultrastructural and genetic investigations strongly favor the sequence of events shown in the diagram. On germination, the haploid spore produces one, two, or sometimes more uninucleate amoeboid or flagellate cells that multiply by cell division, the flagella of the latter being withdrawn before each division. The flagellate cell typically has a pair of flagella, one long and one short, at the anterior end, although two long flagella may sometimes be present. The rudimentary short flagellum is often difficult to detect. Mitosis preceding cytokinesis in the myxamoeba involves breakdown of the nuclear envelope and development of a spindle with polar centrioles. Under conditions unfavorable

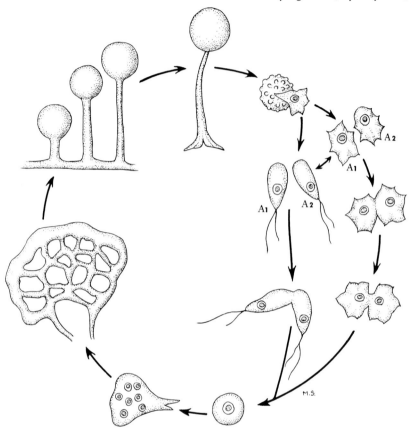

Fig. 114.   Life cycle of a myxomycete. Plasmogamy involving compatible mating types (A₁, A₂) may take place between flagellate cells, myxamoebae, or flagellate cell and myxamoeba. From Olive (1970).

for growth the myxamoebae may become rounded, secrete cell walls, and develop into resting microcysts.

Amoeboid or flagellate cells of compatible mating types eventually fuse to form a binucleate cell in which karyogamy soon occurs. This zygotic cell is the beginning of the plasmodial stage. Its diploid nucleus and all subsequent diploid nuclei in the plasmodium undergo intranuclear division in which the nuclear envelope persists and centrioles are lacking. The nuclei of a plasmodium divide synchronously.

Plasmodia often coalesce to form larger ones. As they wander about, they ingest not only food organisms—in nature these may be bacteria, fungal cells, protozoa, algae, etc.— but also myxamoebae and flagellate cells of its own species, an activity referred to as cannibalism. As the typical plasmodium increases in size a conspicuous rhythmically reversible flow, or shuttle movement, of protoplasm begins within it. In addition to contributing to plasmodial migration, this very active back and forth movement of the protoplasm is undoubtedly involved in maintaining an even distribution of oxygen, nutrients, and metabolites throughout the plasmodium. During mitosis migration, but not streaming, ceases (Rusch, 1968). The plasmodium eventually develops a broad advancing front with a network of main veins leading into it. Under unfavorable growth conditions plasmodia may undergo encystment.

In nature the plasmodium usually develops within the substrate and flows out onto its surface when ready to fruit. In culture, sporulation may occur on the same agar medium on which the plasmodium was isolated, whereas plasmodia of other species may fail to fruit until transferred to nonnutrient media and, in the case of yellow plasmodia, exposed to light. In sporangiate forms, such as the one illustrated in the diagram the plasmodial protoplasm becomes concentrated into a number of small mounds, the sporangial initials, which then proceed to develop into sporangia. As the sporangial protoplast rises on its developing stalk, a portion of the surrounding, protective sheath, often with additional material deposited by the protoplast, remains on the substrate as a thin layer known as the hypothallus. A mitotic nuclear division occurs in the young sporangium, followed by cleavage of the sporangial protoplast into uninucleate spores. A system of sterile threads, known as the capillitium, also frequently develops within the sporocarp. Both light and electron microscopy have fairly conclusively demonstrated that meiosis typically occurs in the maturing spores within the sporangium, with three of the four resultant haploid nuclei degenerating by the time the spore is mature. The mature sporangium is surrounded by the peridium, which in its simplest form appears to be no more than that portion of the sheath surrounding the developing fruiting body but in many myxomycetes has other materials added to it.

A number of slime molds are known to complete their life cycles in single-spore culture. In at least two of these there is evidence of meiosis, indicating that they are homothallic. For the majority of these, however, the information available is insufficent to determine whether they are homothallic or apomictic. With one possible exception, there is no evidence of syngamy in these forms. Observations on some indicate that the initiation of diploidy leading to plasmodium formation may arise by endopolyploidy in the nuclei of the myxamoebae.

On several occasions both homothallism (or perhaps apomyxis) and heterothallism have been reported in the same species of myxomycete. In at least one instance, referred to in Section VIII, the homothallic and heterothallic strains can be crossed (Wheals, 1970).

IV. CLASSIFICATION

The well-illustrated monographic treatment of the myxomycetes by Martin and Alexopoulos (1969) covers the taxonomy of the group in admirable detail, including notes on variants that are encountered in many species. These authors recognize two subdivisions in the division Mycota (Fungi): the Myxomycotina, with the single class Myxomycetes, and the Eumycotina, containing the classes of true fungi. Two subclasses of Myxomycetes are recognized: the exosporous Ceratiomyxomycetidae and the endosporous Myxogastromycetidae. For reasons already noted, the former subclass has been transferred to the Protostelia (Chapter 2), a disposition also accepted in a recent review of Larpent (1972). This leaves the bulk of the group, here treated as subclass Myxogastria, in which the five orders recognized by Martin and Alexopoulos are retained. Other comprehensive and useful taxonomic treatments of the slime molds are the well-illustrated monographs by Lister (1925) and Macbride and Martin (1934).

Because plasmodia are rarely present along with mature fruiting bodies of the same species in nature and because many plasmodia are microscopic in size or otherwise inconspicuous, plasmodial characteristics are generally not very useful for identification purposes. The less conspicuous types of plasmodia are best observed in culture. Alexopoulos (1960a) has identified three major types of plasmodia: (1) the protoplasmodium (Figs, 120, 121, and 144), a microscopic amoeba-like protoplast without veins or reticulations; (2) the aphanoplasmodium (Fig. 155A), thin and inconspicuous and comprised of a reticulum of veins that lead into fanshaped advancing fronts; and (3) the phaneroplasmodium (Figs. 150 and 154A) somewhat like the foregoing in form but thicker and readily visible to the unaided eye. In addition to the above three plasmodial types, several species of the Trichiida have been found to produce plasmodia intermediate between the aphanoplasmodium and the phaneroplasmodium (Alexopoulos, 1960a; McManus, 1962). Shuttle movement of protoplasm occurs in all except the protoplasmodium.

Mature sporocarps are far more useful for identifying myxomycetes. Three main types are recognized. (1) Sporangia are relatively small, sessile or stalked structures of fairly uniform size and shape in any one species. Because many are usually formed by a single plasmodium, sporangia are often clustered. The sporangia are likely to be more scattered in protoplasmodial species, in which the protoplasmodia produce the fruiting bodies singly. (2) The aethalium is a comparatively large hemispherical or cushion-shaped sporocarp derived from all or a major portion of the plasmodium. Aethalia of the same species are variable in size and often in shape. (3) The plasmodiocarp is typically a sessile, irregular, often branched or reticular sporocarp derived from major veins of the plasmodium. Intergradations among all three kinds of sporocarps are known, and more than one type may be found in a single species. The sporangium is probably the most primitive, with the others having evolved from it by the compounding of sporangia into larger sporocarps. Indeed, pseudoaethalia are known in which numerous sporangia are closely compacted into a single mass. In some of these the sporangial walls partially disintegrate, creating a more or less continuous spore-forming area. In some sporocarps considered to be true aethalia, a "pseudocapillitium," thought by some to be remnants of sporangial walls, is present.

Considerable taxonomic value is placed on the color, shape, and size of sporocarps; the presence or absence of stalks and of a sterile structure at the top of the stalk called the columella; and the presence or absence of lime (calcium carbonate) deposits, in either granular or crystalline form, within or on the sporocarps. A quick test for calcium carbonate is to add a drop of dilute acid (e.g., 5% lactic acid) to a sporocarp on a slide and look for the evolvement of carbon dioxide bubbles. Color, size, and wall markings (warts, spines, reticulations, etc.) of spores are also of considerable importance. The ultrastructural studies by Schoknecht and Small (1972) indicate that none of the Myxogastria have entirely smooth spores. The nature of the capillitium, when present, or of the pseudocapillitium is taxonomically useful. The true capillitium is composed of threads that are the product of a directed synthetic process within the sporangial protoplast. Capillitial threads may be solid or hollow, free or interconnected, and smooth or marked with warts, spines, cogs, or spiral bands. In some species they contain deposits of calcium carbonate. Recently, protein electrophoresis has given promise of becoming useful in taxonomy, although it clearly has its limitations (Franke and Berry, 1972; Berry and Franke, 1973).

Unfortunately, the myxomycetes are a complex group taxonomically. Many of the characteristics used in the construction of keys are quite variable, both in nature and in culture. For example, under certain environmental conditions sporocarps that normally have conspicuous lime deposits on their surfaces may have little or none, thus leading the investigator to choose an incorrect branch of the key, which in turn leads to the wrong order or family. The same is often true of sporocarp color and form. For such reasons, there is no existent key to the myxomycetes that is entirely suitable. Recently, Locquin (1967) has prepared a kind of synoptic key to the genera in the form of a stack of cards with punchholes to mark the various characters. Although this method also is not without disadvantages, it at least represents an interesting new approach to the problem of identifying myxomycetes. Probably no final solution will be possible without a much greater knowledge of these organisms and the extent of their variability.

Considering the size of the subclass Myxogastria and that the recent monograph by Martin and Alexopoulos is readily available, we offer here only a key, based largely on theirs, to the five orders recognized by these authors. For reasons presented in Section V,C, the proposal by Ross (1973) that a new subclass of myxomycetes embracing the Stemonitida be recognized is not accepted. The key is followed by brief discussions of the orders and their families and by short descriptions of some exemplary species. Some of the ambiguities regarding such features as spore color, presence or absence of lime deposits, and capillitial characters may be detected in the key.

Key to the Orders of Myxogastria*

Spores in mass pallid to bright colored, sometimes brownish, rarely dull black; lime deposits generally absent
  Capillitium typically present
    Sporocarps minute stalked sporangia not exceeding 1.5 mm in height, usually much smaller; capillitium scanty to well developed, sometimes absent; spores white, light colored, or brown; diploid trophic stage a protoplasmodium...... **A. Echinosteliida**
    Sporocarps sporangiate or plasmodiocarpous, usually larger than the above; capillitium present, typically of sculptured (sometimes smooth), hollow (occasionally solid) threads; spores in mass generally pale to bright colored (yellow, orange, red), rarely dark; diploid trophic stage (in species studied) intermediate between phaneroplasmodium and aphanoplasmodium........................ **B. Trichiida**
    True capillitium absent; pseudocapillitium present or absent; sporocarps various; spores in mass pallid to smoky or yellow brown; diploid trophic stage typically a protoplasmodium or phaneroplasmodium .... **E. Liceida**
Spores in mass usually dark purplish brown to black, rarely pale; lime deposits present or absent
  Lime deposits rarely present and then restricted to hypothallus, base of peridium, stalk, or columella; diploid trophic stage typically an aphanoplasmodium.................. **C. Stemonitida**
  Lime deposits usually present in or on the peridium and/or capillitium; diploid trophic stage typically a phaneroplasmodium .... **D. Physarida**

---

* Capital letters preceding the orders refer to the subsections of Section V in which the orders are discussed.

V. DESCRIPTIONS OF TAXA

A. Echinosteliida

These stalked-sporangiate myxomycetes are among the smallest in the class. The spores have warty to minutely punctate markings. Two families are recognized. The Echinosteliidae contain the single genus *Echinoste-*

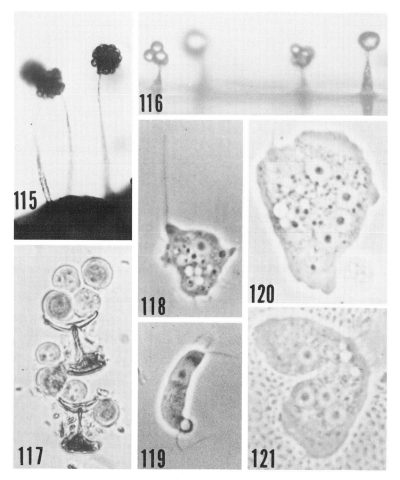

Figs. 115–121. *Echinostelium.* Fig. 115. Sporangia of *E. minutum;* white-spored form, ×217. Figs. 116–121. *E. lunatum.* Fig. 116. Sporangia on agar surface, ×400. Fig. 117. Two 4-spored sporangia with lunate columellas, ×900. Figs. 118 and 119. Uninucleate flagellate cells. Phase contrast ×1260. Figs. 120 and 121. Four-nucleate protoplasmodia. Phase contrast ×1260. Fig. 115 from Olive (1960); Figs. 116–121 from Olive and Stoianovitch (1971c).

lium (Figs. 115–121) with six known species, the largest of which has sporangia not exceeding 0.5 mm in height. A capillitium may be present or absent. The best known members of the genus are a pink-spored form commonly found on bark of coniferous and frondose trees and a white-spored one of common occurrence on moribund plant structures (pods, capsules, etc.), both of which are now classified as *E. minutum* De Bary. They are almost certainly two distinct species (Hung and Olive, 1972a), but since the type collection has not been located there is uncertainty as to which form deserves the specific epithet. Both produce minute sporangia 0.3–0.5 mm tall. Efforts to obtain the pink-spored form in culture have failed, but the white-spored one is readily grown in the laboratory. Haskins (1970) has maintained myxamoebae of the latter in liquid axenic culture. The protoplasmodia of *Echinostelium* resemble multinucleate amoebae (Figs. 120 and 121). They reportedly divide by binary fission (Haskins, 1968), and during sporulation each produces a single stalked sporangium (Figs. 115–117). Single-spore cultures complete the life cycle (Alexopoulos, 1960b; Olive, 1960). Ultrastructural evidence of a meiotic process in the spores (Haskins *et al.*, 1971) indicates that this form is homothallic rather than apomictic.

The smallest and probably the most primitive known myxomycete is *E. lunatum* Olive & Stoian. (1971c), found on dead plant structures in Puerto Rico and North Carolina. It has stalked sporangia less than 50 μ tall with only four to eight, sometimes up to 14, spores (Figs. 116 and 117). A lunate columella occurs at the top of the stalk. The peridium is no more than a thin, fragile sheath that collapses around the spores. A protostelioid disc at the base of the tubular stalk probably represents a primitive hypothallus. Of particular interest are the spore wall markings as seen in thin sections (Hung and Olive, 1972a; see Section VII). These closely resemble the spore wall ornamentations found in the flagellate protostelid *Cavostelium bisporum* (Chapter 2). Flagellate cells typically have one long anterior flagellum and an inconspicuous, short, recurved one, although some have two long flagella and others have supernumerary ones unassociated with the nucleus (Figs. 118 and 119). The amoeba-like protoplasmodia contain only two to six nuclei. These and other characteristics of *E. lunatum* indicate that it may be more closely related to the protostelids than is any other known myxomycete.

A second family, the Clastodermidae, contains only three minute sporangiate species with brown spores, two belonging to the genus *Clastoderma* and one to *Barbeyella*. The family was recently described by Alexopoulos and Brooks (1971), the two genera having been transferred here from the order Stemonitida.

Echinosteliopsis oligospora Reinhardt and Olive (1966), an unusual mycetozoan with many of the characteristics of *Echinostelium*, was

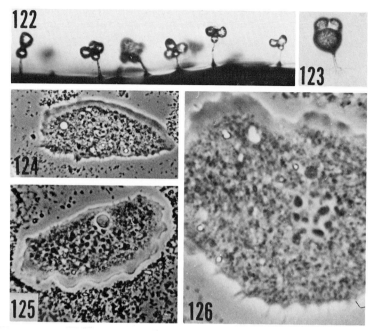

Figs. 122–126. *Echinosteliopsis oligospora.* Fig. 122. Sporangia on agar surface, ×195. Fig. 123. Sporangium with large and small spores, ×195. Figs. 124–126. Amoeboid trophic stage. Note multinucleolate nuclei. Phase contrast ×435, ×25, and ×1326, respectively. From Reinhardt (1968b).

placed by Olive (1970) in a new family, Echinosteliopsidae, and order, Echinosteliopsida, of uncertain affinities. *Echinosteliopsis oligospora,* which is of fairly wide occurrence on dead plant structures, produces minute, mostly 4- to 8-spored *Echinostelium*-like stalked sporangia with smooth spores (Figs. 122 and 123). The trophic stage has one to several nuclei and typically resembles a protoplasmodium (Figs. 124–126), although limited reticulate protoplasts with larger numbers of nuclei may also develop in culture. Filose pseudopodia are characteristic. No sexual process has been observed. The species differs strikingly from *Echinostelium* in lacking a flagellate stage and in having nuclei with many peripheral nucleoli instead of a single central one (Fig. 126). It has been studied in some detail by Reinhardt (1968b). Whether it evolved independently of other mycetozoans or from *Echinostelium* remains a question.

Bursulla crystallina Sorokin (1876), found on horse dung in Russia, has been placed tentatively in the Echinosteliopsidae (Olive, 1970), although its true relationships remain uncertain. It produces minute stalked sporangia with eight unwalled cells. The latter, on leaving the

112   4. Myxogastria (Myxomycetes)

sporangium, develop filose pseudopodia but no flagella. The report of
plasmogamy in this species needs further study. *Bursulla crystallina* appears to be known only from the original collection.

B. Trichiida

The sporocarps of this order are usually stalked or sessile sporangia
or plasmodiocarps. The spores are light or bright colored, often yellow
in mass, and the capillitial threads are typically well developed, solid
or tubular, elastic, and sculptured in various ways, frequently with spiral
ridges (Fig. 136). The spore walls are marked with delicate spines or
warts or with reticulations. Two families are recognized.

The Dianemidae, with four genera, have a capillitium of solid threads
that are attached to the base and often the sides of the sporangial wall
and that do not form a network.

The Trichiidae are the larger family, containing ten genera and numerous species, including some of the commoner and better known
myxomycetes. The tubular, hygroscopic capillitial threads are generally
ornamented with spirals, cogs, rings, or spines and are often united into
a network. They occur free in the sporangium or attached to its base.
Perichaena has a double peridium and capillitial threads with markings
other than spiral ridges. The brownish sporangia of *P. depressa* Libert,
a cosmopolitan species on dead bark and wood, are sessile and flattened.
usually several sided, and open by means of a lid (Fig. 127). *Arcyria*

Figs. 127–131. Myxomycete sporocarps. Fig. 127. *Perichaena depressa,* sessile
sporangia. Note sporangial lids, ×5.5. Fig. 128. *Arcyria cinerea,* stalked sporangia,
×5.5. Fig. 129. *Metatrichia vesparium,* clustered sporangia with common stalks,
×7.3. Fig. 130. *Trichia varia,* sessile sporangia, ×7.3. Fig. 131. *Hemitrichia serpula,*
reticular plasmodiocarp, ×2.

has stalked, globose to nearly cylindric sporangia, the peridium of which disappears early except at the base, where it persists as a cup to which the variously ornamented, elastic capillitial threads remain attached. The spores in mass have a variety of colors—yellow, pinkish, reddish, ochraceous, etc. *Arcyria cinerea* (Bull.) Pers., with gray to ochraceous sporangia, is common on dead wood and humus (Fig. 128). *Metatrichia vesparium* (Batsch) Nann.-Brem., one of the most common myxomycetes on rotting wood, produces dark maroon to nearly black sporangia clustered in an adherent group at the top of a common reddish stalk (Fig. 129). Each sporangium opens by a lid, followed by emergence of the reddish, elastic capillitium, the threads of which are ornamented with spiral bands. The spores are brownish red in mass. *Trichia*, a genus of 14 species, produces clusters of many small stalked or sessile sporangia, generally with yellow capillitium and spores. The capillitium is composed of free, simple or sparsely branched threads (Fig. 136). *Trichia varia* (Pers.) Pers. (Fig. 130) produces yellowish sporangia that are sessile or nearly so and often show a tendency to become plasmodiocarpous. One of the most conspicuous plasmodiocarpous myxomycetes is *Hemitrichia serpula* (Scop.) Rost., on dead wood and humus (Fig. 131). The yellowish reticular fruiting bodies may be up to several inches in length. Capillitium and spores are bright yellow. The genus, which contains mostly sporangiate species, may be distinguished from *Trichia* on the basis that the capillitial threads occur in a network.

C. Stemonitida

The single family Stemonitidae, with its 13 genera, is recognized. The spores are purplish brown to ferruginous or black in mass, and their walls are ornamented with warts, spines, or reticulations. The capillitium is most commonly a network of dark smooth threads. Lime deposits are generally absent from the sporangia, except in the genus *Diachea*, in which the stalk and columella (when present) are predominantly of calcium carbonate. *Diachea leucopodia* (Bull.) Rost., which is common on dead leaves and sticks, has dark irridescent, cylindrical, blackish-spored sporangia with white stalks (Fig. 132). The stalk is continuous with a white columella within the sporangium. One of the best known genera of slime molds is *Stemonitis*, comprised of about 16 species that occur on dead wood and humus. The long, narrow, brown sporangia are borne in clusters on slender glossy black stalks (Fig. 133). The stalk is continuous with a black columella that extends through the center of the sporangium and from which the black capillitial threads diverge, branch, and anastomose peripherally to form a surface

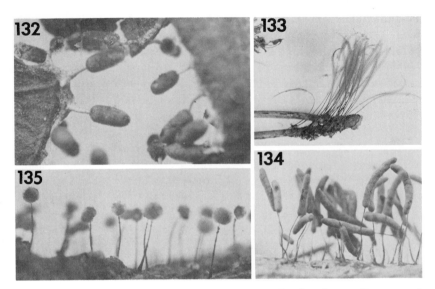

Figs. 132–135. Myxomycete sporocarps. Fig. 132. *Diachea leucopodia,* sporangia with white lime-impregnated stalks, ×10. Fig. 133. *Stemonitis axifera* sporangia after most of the spores have been shed, ×2.7. Fig. 134. *Comatricha typhoides,* stalked cylindrical sporangia, ×7. Fig. 135. *Lamproderma arcyrionema,* stalked globose sporangia, ×17.

net (Fig. 137). *Comatricha,* with globose to cylindrical sporangia, appears to be closely related to *Stemonitis,* from which it is distinguished by the absence of a peripheral capillitial net. *Comatricha typhoides* (Bull.) Rost. (Fig. 134), on rotting wood, produces cylindrical gray sporangia that resemble those of *Stemonitis,* causing some investigators to classify it in that genus. *Lamproderma,* which has globose to subcylindrical stalked or sessile sporangia, may be distinguished by the tendency of the capillitium to arise only from the upper end of the columella, and also by its persistent peridium. *Lamproderma arcyrionema* Rost., on dead leaves and wood, forms stalked globose sporangia that are silvery gray to bronze colored (Fig. 135).

Those interested in pursuing further the classification of this group should refer to the papers by Nannenga-Bremekamp (1967) and Ing and Nannenga-Bremekamp (1967), who propose the recognition of new genera, a new family, and other revisions in the classification of the Stemonitida, and by Ross (1973), who proposes elevation of the order to a new subclass (Stemonitomycetidae) on a par with the remainder of the myxomycetes, a disposition recently accepted by Alexopoulos (1973). Ross's decision is based primarily on the presence of a supraplasmodial hypothallus and a stalk formed inside the culminating sporangial protoplast in the Stemonitida, as opposed to the absence of such a hypothallus and passage of the sporangial protoplast through the interior of the stalk in other myxomycetes. This proposal seems premature for several reasons. First, there is no evidence that these features, considered unique to the Stemonitida, are characteristic of the majority of its genera. Second, there is insufficient information available to permit the claim that there are only two types of sporangial development in the myxomycetes. In fact, it is probable that a still different type of development, that characteristic of protostelids, occurs in *Echinostelium lunatum* (and possibly in some other species of the genus). Third, in no other subclass of eumycetozoans has such emphasis been placed on mode of stalk development, although significant differences also exist with respect to this feature within the Protostelia and Dictyostelia. Under the circumstances, the subclass newly proposed by Ross is considered unacceptable without additional supporting information.

D. Physarida

Like the foregoing order, the Physarida have spores that are dark colored in mass. They are further distinguished by having lime deposits in the peridium, capillitium, or stalk, but never in the stalk alone. The capillitium is composed of branching threads or tubules, often with expanded nodes containing calcium carbonate (Fig. 138). However, the virtual absence of lime from sporocarps of some species growing under certain conditions in the laboratory or in nature can cause difficulties in the use of most keys. The sporocarps are most often sporangiate but are sometimes plasmodiocarpous. The spore walls are warty, spiny, or irregularly verrucose to reticulate. Two large families are recognized.

The Physaridae, with eight genera and a large number of species, is distinguished by the calcareous deposits that are present in the capillitium and often in other parts of the sporocarp. These lime deposits are of a granular nature. Intergradations between genera and variations among species add to identification problems in this family as well as the next. *Physarum*, with about 85 species, is the largest genus of myxomycetes. The sporocarps are most often sporangiate but are plasmodiocarpous in some species. The genus shows such a wide range of characters that some authorities have attempted to subdivide it into several genera; however, Martin and Alexopoulos (1969) have preferred to retain the single genus. The best known species, *P. polycephalum* Schw. (Fig. 139), is common on dead wood and other substrates. It has long been a favorite of laboratory investigators by virtue of its conspicuous yellow phaneroplasmodia that grow and fruit well in culture. The grayish, stalked sporangia are typically gyrose and lobed, appearing of a semicompound nature. *Leocarpus* is recognized by its black-spored sporangia with brittle three-layered peridium. *Leocarpus fragilis* (Dicks.) Rost. (Fig. 140) produces shiny yellowish or brownish egg-shaped sporangia, usually on short stalks. The gregarious sporangia are commonly found on dead twigs, rotting wood, and humus. *Fuligo* is an aethalioid genus with both a pseudocapillitium and a true capillitium containing lime nodes. The former may represent wall remnants of individual sporangia from which the aethalium probably evolved. *Fuligo septica* (L.) Wiggers sometimes produces aethalia about a foot in length, though most often they are from one to a few inches broad. The pulvinate fruiting bodies are typically ochraceous until the relatively thick calcareous covering layer breaks away and reveals the black spores (Fig. 141). This is a common and widespread species on rotting wood, humus, mulch piles, manure, etc.

The Didymiidae, with six genera, differ from the foregoing family

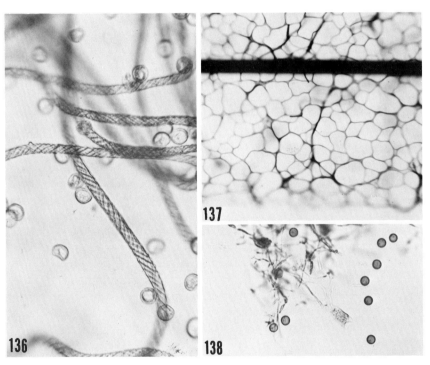

Figs. 136–138. Myxomycete capillitia. Fig. 136. *Trichia* sp., spores and capillitial threads with spiral ornamentations, ×480. Fig. 137. *Stemonitis* sp., columella and capillitial network, ×150. Fig. 138. *Physarum polycephalum*, spores and capillitial threads, showing a calcareous node, ×225.

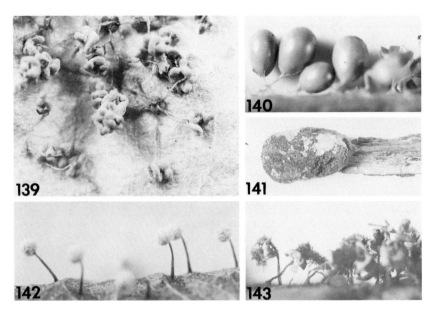

Figs. 139–143.   Myxomycete sporocarps. Fig. 139. *Physarum polycephalum*, stalked gyrose sporangia, ×6.2. Fig. 140. *Leocarpus fragilis*, egg-shaped sporangia, ×15.5. Fig. 141. *Fuligo septica*, aethalium, ×0.6. Fig. 142. *Didymium iridis*, stalked globose sporangia, ×15.5. Fig. 143. *Diderma floriforme*, stalked sporangia with reflexed peridial segments, ×7.

in having a capillitium that is usually devoid of lime deposits (although exceptions occur), whereas calcareous deposits are often present elsewhere in the sporocarp. They are often of a crystalline nature. *Didymium*, a genus of about 30 species, has sporangiate or plasmodiocarpous fruiting bodies with calcareous crystals covering the peridium. *Didymium iridis* (Ditmar) Fr. (Fig. 142), a common species studied frequently in the laboratory, forms small, stalked, globose sporangia on humus and dead wood. *Diderma*, a genus of about equal size, differs from *Didymium* in having amorphous, granular lime deposits on the peridium. *Diderma floriforme* (Bull.) Pers. (Fig. 143) produces stalked sporangia, the peridium of which folds back in petal-like lobes exposing the calcareous clavate columella, capillitium, and dark spores.

E. Liceida

This order contains species that show a wide range in form of sporocarp (sporangia, plasmodiocarps, aethalia) (Figs. 144–149). There are pseudoaethalioid species, the fruiting bodies of which show varying de-

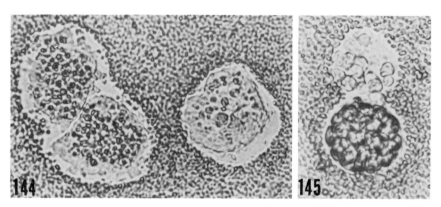

Figs. 144 and 145. *Licea* sp. Fig. 144. Protoplasmodia feeding on yeast cells, ×540. Fig. 145. Minute sessile sporangia, ×540.

grees of compounding of sporangia into single large sporocarps, suggesting the probable phylogenetic sequence of events leading to the evolution of aethalioid forms from sporangiate ones. A true capillitium is absent, although a pseudocapillitium may be present. The spores in mass vary from pale to yellowish brown but not purplish brown. Spore walls are ornamented with minute spines or warts or with reticulations. Ultrastructural studies are needed to determine whether several reportedly smooth-spored species may not have minute markings. Three families are recognized by Martin and Alexopoulos.

The Cribrariidae, with three genera, contain mostly sporangiate forms in which anastomosing ridges within the peridium persist as a net around the spore mass. Small granules, called "dictydine granules," are found around the spores and other structures in the sporangium. *Cribraria* (Fig. 146) has a delicate peridial net without prominent longitudinal strands, whereas *Dictydium,* as exemplified by *D. cancellatum* (Batsch) Macbr. (Fig. 147), does have such strands. The latter is common and widespread on dead wood.

The single genus *Licea*, containing 19 species, comprises the family Liceidae. The fruiting bodies are either plasmodiocarps or sessile to stipitate sporangia lacking capillitium, pseudocapillitium, and dictydine granules. The smallest species of *Licea* produce minute sporangia, derived from protoplasmodia (Fig. 144), that do not exceed 150 μ in diameter. These sometimes consist of little more than tiny mounds of spores covered by a thin transparent peridium (Fig. 145). *Licea biforis* Morgan has been cultured from spore to spore in the laboratory (Wollman and Alexopoulos, 1967).

The Reticulariidae have fruiting bodies that range from sporangiate

Figs. 146–149. Myxomycete sporocarps. Fig. 146. *Cribraria* sp., stalked sporangia showing nets with dictydine granules, ×15. Fig. 147. *Dictydium cancellatum;* stalked sporangia showing nets with prominent longitudinal ribs, ×9. Fig. 148. *Dictydiaethalium plumbeum;* pseudoaethalium comprised of many partially fused sporangia and seated on a pale hypothallus, ×3. Fig. 149. *Lycogala epidendrum,* aethalia, ×1.8.

to pseudoaethalioid or truly aethalioid and usually contain a pseudocapillitium. Four genera are recognized. *Dictydiaethalium* produces flattened cushionlike pseudoaethalia that well exemplify the intermediate condition between closely appressed sporangia and a true aethalium. Portions of the sporangial walls disappear during maturation, and the remaining wall segments are referred to as "pseudocapillitium." There are only two known species, of which only *D. plumbeum* (Schum.) Rost. (Fig. 148), on dead wood, is relatively common and widespread. The yellowish brown to olivaceous sporocarps are situated on a conspicuous hypothallus. *Lycogala* has pulvinate or somewhat conical, brownish aethalia that may at first be mistaken for small puffballs. A pseudocapillitium of simple to branched, mostly hyaline tubes is present, and the spores are pinkish to ochraceous in mass. *Lycogala epidendrum* (L.) Fr. (Fig. 149), on dead logs, is one of the most common myxomycetes. It is doubtful whether the structures referred to as pseudocapillitia in *Dictydiaethalium* and *Lycogala* are homologous, which suggests that two different terms are needed here.

VI. DETAILS OF THE LIFE CYCLE

A. The Trophic Haplophase

The haploid trophic stage of a slime mold begins with the naked uninucleate protoplasts that emerge from germinating spores. The spores of various species show conspicuous differences in longevity, some apparently remaining viable for less than a year, whereas those of *Lycogala flavofuscum* (Ehrenb.) Rost. (Elliott, 1949) and *Hemitrichia clavata* (Pers.) Rost. (Erbisch, 1964) have been reported to germinate after dry storage for 68 and 75 years, respectively. The spores of most myxomycetes can probably survive under suitable storage conditions for at least several years.

The spores of a number of species do not germinate well in the laboratory without prior treatment. Elliott has succeeded in germinating spores of 57 different myxomycetes by first centrifuging them in a 1% solution of a wetting agent (sodium glycholate or sodium taurocholate, or equal mixtures of the two). In such species as *Echinostelium minutum* and *Reticularia lycoperdon* Bull., germination may occur within half an hour; whereas in such species as *Dictydium cancellatum* as many as 18 days may be required before the start of germination.

Although mature spores of at least a few myxomycetes may contain more than one nucleus, usually only a single nucleus is present. During germination, however, it is not uncommon for more than one uninucleate protoplast to emerge from the spore, and the number may be as high as eight in *Badhamia affinis* Rost. This is generally thought to result from mitosis followed by cytokinesis in the spore prior to germination. The protoplasts emerge through a pore or slit in the spore wall. If adequate moisture is present they usually exit as flagellate cells or develop flagella soon thereafter. If there is insufficient moisture the protoplasts remain amoeboid and nonflagellate. Myxamoebae readily become flagellate in the presence of water.

Each flagellate cell has one, or sometimes two, long anterior flagella (Fig. 118). Extra flagella may also be present (Fig. 119). Even when only a single long flagellum is apparent, staining or phase-contrast microscopy generally reveals a second, short, and reflexed flagellum at the base of the longer one (Elliott, 1949; Locquin, 1949), and electron microscopy has clearly shown that the basal flagellar apparatus contains two kinetosomes (Section VII). It must be concluded, therefore, that myxomycetes, like flagellate protostelids, are basically biflagellate. It is not uncommon for a slender filopodial extension resembling a flagellum to appear repeatedly at the anterior end of the flagellate cell and move backwards, finally being withdrawn at the posterior end. This structure, termed a "pseudoflagellum," is quite possibly involved in cell movement. Pseudoflagella, which lack the internal structure of true flagella, have also been observed in flagellate protostelids.

Both flagellate and amoeboid cells feed by ingesting cells of the food organism (e.g., bacteria, yeasts) and both undergo cell division. Just before the flagellate cell divides it withdraws its flagella, a development probably associated with the interchangeability of kinetosomes and centrioles and with the requirement of the latter in nuclear division. Conclusions by earlier investigators (e.g., Wilson and Cadman, 1928) that this division is of the extranuclear or astral type, with the nuclear envelope breaking down at late prophase, have been repeatedly confirmed (e.g., Koevenig, 1964; Kerr, 1967; Ross, 1967a). An ultrastructural study by Aldrich (1969) leaves no doubt as to the presence of centrioles at the spindle poles and the breakdown of the nuclear envelope. The process of constriction into two daughter cells is very similar to that in protostelids and dictyostelids. The nucleolus disappears in late prophase and first reappears in each daughter nucleus as several small pronucleoli that fuse into a single nucleolus (Koevenig, 1964; Kerr, 1967). The daughter cells may quickly develop flagella. Ross (1967d) has found that some of the flagellate cells of a *Didymium iridis* isolate studied by him produce more than one set of flagella, in which case only one set is associated with the nucleus—a condition found to be common in some protostelids. Similar behavior has been reported in *Echinostelium lunatum* (Olive and Stoianovitch, 1971c) (Fig. 119).

In cultures of *Didymium iridis* and *Badhamia curtisii* that had been maintained axenically for several years and had lost the ability to form plasmodia, Ross (1968) discovered a remarkable amount of heteroploidy. The chromosome number in dividing amoebae of the former species ranged from 20 to 300. As Kerr (1967) had found in a related investigation of *D. nigripes* (Link) Fr., Ross observed that regardless of the ploidy level in the amoeboid cell the nuclear envelope broke down during mitosis, and the division therefore did not resemble the intranuclear division of diploid plasmodial nuclei. These and other studies reveal that variations in ploidy are common in myxomycetes and probably result in large part from autopolyploidy.

Under unfavorable growth conditions amoeboid cells often become quiescent and spherical and secrete relatively thin walls around themselves, thus becoming microcysts. When flagellate cells encyst the flagella are first withdrawn. Nuclei of microcysts remain in the G_1 (unreplicated) phase. On being transferred to fresh growth media, these resting cells germinate and produce the haploid trophic stage again.

B. Syngamy and Initiation of the Diplophase

Syngamy in heterothallic myxomycetes has been found to occur between pairs of flagellate cells, between pairs of myxamoebae, or between flagellate cell and myxamoeba. The genetics of compatibility in heterothallic slime molds is discussed in Section VIII. No morphological sexual differences between gametes have been found. Several investigators (e.g., Ross, 1967a–c; Ross *et al.*, 1973) have noted that a period of time is required for haploid cells to become physiologically conditioned before they can function as sexual cells. For example, cell fusions in *Perichaena vermicularis* (Schw.) Rost. have been found to occur 12–18 hours after spore germination (Ross, 1967c). When myxamoebae of compatible mating types of *Didymium iridis* are first mixed together they show no attraction for each other, but after about 10–12 hours syngamy proceeds rapidly (Ross, 1967a). Ross suggests that inducers secreted by compatible myxamoebae after they have come into close proximity or into actual contact may be required to trigger cell fusions by physiologically conditioning the cell membranes.

Although there has been some disagreement in the past over where karyogamy occurs in the life cycle, studies by several recent investigators (e.g., Koevenig, 1961; Ross, 1967a) have confirmed the conclusion of Jahn (1911) that it occurs immediately after syngamy. Koevenig, using phase-contrast cinematography, beautifully illustrated both syngamy and karyogamy in *Physarum gyrosum* Rost. In *Didymium iridis* karyogamy usually occurs 4–20 minutes after syngamy, followed by nucleolar fusion 2–3 minutes later (Ross, 1967a).

As noted in Section III, some species of myxomycetes complete their life cycles in single-spore culture, and for most of these it is not known whether they are homothallic or simply apomictic. Von Stosch (1937) has considered such species apogamous, with gametes but not meiosis eliminated from the life cycle. A strain of *D. nigripes* has been described that forms plasmodia reportedly without prior syngamy or karyogamy (Kerr, 1967, 1968, 1970; Kerr and Kerr, 1967). Ploidy was found to vary among the nuclei of both myxamoebae and plasmodia. There was also evidence that haploid plasmodia may develop from haploid amoebae. Ross (1966) observed similar variations in ploidy in several other myxomycetes and has suggested that a diploid myxamoeba may produce a plasmodium without the requirement of syngamy, or that a haploid plasmodium may sporulate without meiosis. The difficulty of determining whether diploidy in slime molds without mating type systems arises through syngamy or through endopolyploidy is obvious.

In a study of *D. nigripes*, Kerr (1968) concluded that myxamoebae may fuse without giving rise to plasmodia and that cell division continues in the manner characteristic of amoeboid cells. Such fusions appear to resemble nonsexual cell fusions in protostelids and cellular slime molds. Kerr also reported that some larger cells, probably polyploid, may undergo tripolar divisions.

The recent discovery that synaptonemal complexes occur in the spores of *Physarum pusillum* (B. & C.) G. Lister and *Echinostelium minutum*—two species that complete their life cycles in single-spore culture—indicates that these two species are truly homothallic. Apparently the only homothallic species in which both syngamy and karyogamy have been confirmed is *Didymium squamulosum* (Alb. & Schw.) Fr. (Ross, 1957a).

Two types of cell fusions that may lead to plasmodial development have been reported in heterothallic *D. iridis* (Ross and Cummings, 1970). When compatible clones of myxamoebae are grown separately for 60–68 hours of exponential growth and then mixed together, normal syngamy occurs after an induction period of 5–6 hours. However, if the same clones are grown for 44–52 hours and mixed, the myxamoebae quickly begin to fuse in twos, threes, fours, fives, etc., without the requirement of an induction period. This is followed by the fusion of nuclei in different combinations—two haploid, two diploid, haploid and diploid, etc. These developments result in syncytia containing nuclei varying in size and ploidy. Some of the fusion products develop into plasmodia that fruit normally and show normal segregation of mating types in the sporangia. It is concluded that once the barrier to cell membrane fusion disappears there is no barrier limiting the fusion of compatible nuclei.

C. Trophic Diplophase: The Plasmodium

1. Gross Morphology and Cytology

The major trophic stage of a myxomycete is its plasmodium. The most frequently investigated plasmodium is that of *Physarum polycephalum*, now maintained in axenic culture in a number of laboratories. It is commonly observed in nature and in the laboratory as a bright yellow, flattened, fan-shaped mass with major and lesser veins forming a reticulum leading into a continuous, advancing front (Fig. 150). The color of this and certain other phaneroplasmodia is known to vary with pH and pigmentation of food organisms ingested.

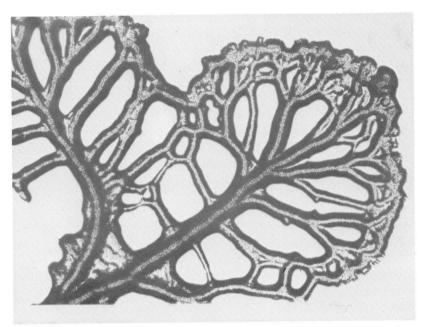

Fig. 150. Phaneroplasmodium of *Physarum polycephalum*, ca. ×43. Culture supplied by D. G. Humm.

Whereas the plasmodia of heterothallic species originate via syngamy and karyogamy, a different type of fusion occurs between plasmodia (Section VIII,B). In such species as *P. polycephalum* and *D. iridis* uninucleate to plurinucleate microplasmodia or multinucleate macroplasmodia may coalesce to form larger ones. However, single plasmodia may develop without coalescence into large plasmodia that fruit. When thousands of microplasmodia of *P. polycephalum* grown axenically in liquid shake cultures are placed in a concentrated drop on a culture plate, they coalesce within a few hours into a single, relatively large, veined, migrating plasmodium.

Nuclear division in the plasmodium is distinctly different from mitosis in the myxamoeba in that the spindles lack polar centrioles and are intranuclear (Howard, 1932; Koevenig and Jackson, 1966). In addition the divisions are synchronous. Plasmodial mitoses in *P. polycephalum* are estimated to occur at intervals of from 8 hours (Chin *et al.*, 1972) to 15.4 hours (Guttes *et al.*, 1959). When plasmodia in different stages of the nuclear division cycle coalesce, the next mitosis is asynchronous but subsequent mitoses become synchronous. Mitotic divisions in plasmodia of this species have been clearly illustrated by Koevenig and

Jackson (1966), who report a diploid chromosome number of about 56, and by Koevenig (1961) in *P. gyrosum*. Variations in chromosome numbers among plasmodial nuclei within the same species, or even in the same isolate, probably indicate varying degrees of ploidy. Chromosome counts in plasmodial nuclei of *P. polycephalum* have ranged from 20 (Guttes *et al.*, 1961) to 90 (Ross, 1961) or occasionally to more than 100. This very nearly coincides with the range commonly reported for myxomycetes in general. There is evidence that extended periods of laboratory culture are likely to result in wide ploidy ranges within a single isolate (Ross, 1966). Some ploidy changes occur as a result of incomplete nuclear division (Kerr, 1967). An improved aceto-orcein smear technique for making chromosome counts has recently been described (Mohberg *et al.*, 1973).

During mitotic prophase in the plasmodium the nucleolus degenerates, and each new nucleolus is reconstituted by the fusion of several pronucleoli that appear in the daughter nuclei. Reports of two to five nucleoli in plasmodial nuclei probably refer to stages before the reconstitution of the single larger nucleolus (Rusch, 1968).

Shuttle movement of the protoplasm begins in the plasmodium while it is still microscopic in size. In *Perichaena vermicularis* it begins at about the 20-nucleate stage (Ross, 1967b). The description of plasmodial development in *P. polycephalum* by Camp (1937) remains one of the most detailed available. The advancing front of the plasmodium is usually a continuous sheet, with a reticulum of veins leading into it. The veins originate from thin hyaloplasmic lobes, often with filose pseudopodia, that develop along the advancing margin. These soon become granular in appearance as they elongate and anastomose to form additional reticulate veins of the plasmodium. Veins may also originate as outgrowths from perforations in other parts of the plasmodium. Camp describes the veins as being composed of a gel-like peripheral layer through which the sol, or more liquid component, of the plasmodium flows. The entire plasmodium is covered by a thin slime sheath, which, as the plasmodium continues its migration, remains behind in collapsed form on the substrate. It resembles in some ways the slime sheath of *Dictyostelium discoideum*, which is deposited on the substrate behind the migrating pseudoplasmodium.

The sheath covering the plasmodium is thinnest at the advancing margin. A recent report (Simon and Henney, 1970) that it is comprised of a glycoprotein containing galactose has been modified by McCormick *et al.* (1970a), who claim that it is a sulfated galactose polymer with traces of rhamnose. The latter investigators suggest that, among other possible functions, the sheath serves as a protective agent and possibly as an ion-exchange material. It is clearly not comparable to the hyphal walls of fungi, as has sometimes been claimed (Martin, 1960). Although the plasmodium is a single multinucleate protoplast, pieces can be cut out of it and subcultured without permanent damage, especially in the case of a phaneroplasmodium. Camp (1937) and others have noted that mechanical injury, such as pressure or puncture from a needle, causes temporary gelation of protoplasm and cessation of protoplasmic flow in the affected area. A quick puncture may first result in expulsion of a mass of plasmasol, demonstrating that the plasmodial contents are under pressure (Andresen and Pollock, 1952). Resorption of small severed veins or of protoplasm expelled from larger cut veins may also occur. These remarkable regenerative powers are especially characteristic of actively growing plasmodia.

Indira (1964) reported an activity by plasmodia of *Stemonitis herbatica* Peck, *Physarum compressum* Alb. & Schw., and *Arcyria cinerea* (Bull.) Pers. that is either unusual or seldom noticed. On transfer to sterile water the plasmodia pinched off flagellate cells, the cells sometimes becoming flagellate before complete detachment. Small plasmodia that later appeared were thought to arise from these cells. There is no information on whether the nuclei of the flagellate cells were haploid or diploid. In the absence of centrioles in plasmodia, the origin of the flagellar apparatus of the planonts is also unclear. Ross and Cummings (1967) described the origin of uninucleate amoeboid cells from plasmodia of an unidentified *Physarum* but did not determine their ploidy or subsequent development. In both of these examples the smaller cells appear to be produced by plasmotomy. These developments need to be studied in greater detail.

2. Mitotic Cycle in the Plasmodium

Large phaneroplasmodia of such species as *P. polycephalum* contain millions of nuclei at a density of about 800,000 per cubic millimeter (Andresen and Pollock, 1952), and all these nuclei divide more or less synchronously at regular intervals. It is therefore not surprising that an increasing number of investigators are turning to the myxomycetes for information on the synthesis of DNA, RNA, and proteins at various stages in the mitotic cycle, including the synthesis of substances that instigate and coordinate mitosis. A group at the University of Wisconsin (Rusch, 1968; Brewer and Rusch, 1968; Cummins and Rusch, 1968) have been particularly active in this field; their results on this and related subjects have been reviewed by Cummins (1969). The considerable amount of information that has accumulated on macromolecular events of cell and nuclear cycles during plasmodial development have been reviewed by Schiebel (1973). Only a few major aspects of these investigations are considered here.

With the aid of radioactive isotopes it has been determined that DNA replication is synchronized during the nuclear cycle, as is RNA transcription. There is essentially no G_1 phase (the period between mitosis and DNA synthesis), and DNA synthesis starts immediately after telophase. It reaches its full rate within 10 minutes and continues at a high rate for about 2 hours, dropping markedly during the next hour. RNA synthesis has two peaks, one about 2.5 hours after telophase and the other about 2 hours before the next metaphase. The messenger RNA and protein involved in the synchronization of mitosis appear to be synthesized during the latter half of the second period of RNA synthesis, the last protein essential to mitosis being transcribed only 15 minutes before metaphase. A heat shock delivered to the plasmodium during this second period inhibits mitosis, apparently by destroying a heat-sensitive mitotic inducer, and results in an endoploidal replication of DNA. This is another possible mechanism for the origin of variations in ploidy in plasmodia.

From isolated plasmodial nuclei Jockusch et al. (1971, 1973) extracted an actin-like protein that was localized primarily in the nucleoli and resembled cytoplasmic actin but was synthesized only in the period preceding mitosis. It can be made to form thick fibers similar to myosin from striated muscle. They suggest that the spindle microtubules may be composed of actin and that the presence of actin in the nuclei may be associated with nuclear division in plasmodia. The basic nuclear proteins of myxomycetes are similar to the histones of most other eukaryotes but are unlike those of fungi that have so far been analyzed (Coukell and Walker, 1973; Jockusch, 1973).

The details of mitosis in the plasmodium are reserved for Section VII,B, which deals with the ultrastructure of the plasmodium.

3. Protoplasmic Streaming and Migration of the Plasmodium

Protoplasmic streaming, occurring in periodically alternating directions, continues regardless of whether the plasmodium is migrating or stationary. Migration ceases when the plasmodium reaches an adequate food supply, as well as during mitosis and just prior to fruiting. The duration of protoplasmic flow in either direction is variable, depending in large part on culture conditions and vigor of the plasmodium. In *P. polycephalum* this period is often around 1 minute but may vary from 0.5 to 10 minutes (Jahn, 1964). According to Kamiya (1950), locomotion of the plasmodium is controlled less by the duration of flow in one direction than by the volume of protoplasm transported. Cummins and Rusch (1968) reported that in well-fed nonmigrating plasmodia, alternate periods of flow are of about equal duration, but in unfed migrating plasmodia the period of flow toward the advancing front tends to be of somewhat longer duration than the reverse period of flow.

A better understanding of protoplasmic streaming in myxomycetes may be gained by reference to certain gross features of plasmodial structure, to be discussed here, and to the fine structure of the plasmodium, discussed in Section VII,B. The outermost layer encasing the plasmodium is the slime sheath. This is followed by the plasmalemma of the protoplast itself. In the area of the veins a gel layer of protoplasm lies just within the plasma membrane, and the more liquid (sol) part of the protoplasm flows through the central area of the vein. Camp (1937) and later investigators have noted that sol and gel components are not essentially different from each other in makeup but differ primarily in degree of fluidity. One zone merges into the other insensibly and the sol and gel phases are readily interconvertible.

Several interesting theories have been proposed to explain protoplasmic streaming in myxomycete plasmodia, more detailed discussions of which can be found in Gray and Alexopoulos (1968). Stewart and Stewart (1959) have proposed the "diffusion drag force hypothesis," which states essentially that some particular compound(s) synthesized by the protoplast diffuses down a concentration gradient, in the process of which some of the momentum is transferred to adjacent molecules, which are then dragged along with them. The theory has not gained wide acceptance.

A second hypothesis—one which is gaining increasing support from physiological and ultrastructural studies—is the "contraction-hydraulic theory" of Jahn and co-workers (Jahn, 1964; Jahn et al., 1964), who have reviewed the current concepts of protoplasmic flow in much greater detail than is feasible here. Most of their findings support those of Camp, who prophetically suggested that the eventual explanation of protoplasmic flow may have much in common with muscular contraction. Loewy (1952) was the first to isolate from the plasmodium an actomyosin-like protein that resembles the actomyosin of animal muscle in its ability to contract in the presence of adenosine triphosphate (ATP). This is consistent with the observation that plasmodia contract at the points at which ATP is applied (Nakajima, 1964). In addition, ultrastructural evidence to be discussed in Section VII,B strongly implicates actin-complexed fibrils in the contractile process. Supporting evidence was also presented by Hatano and Oosawa (1962, 1964), who isolated from plasmodia of P. polycephalum an actin-like protein that formed a complex with muscle myosin that was similar to muscle actomyosin. Oosawa *et al.* (1966) obtained a similar complex between plasmodial myosin and muscle actin. Nachmias (1972) separated the myosin from plasmodial actomyosin and precipitated it from a potassium chloride solution. The precipitate consisted of aggregated, short, bipolar filaments that were similar to those observed in fixed and sectioned material. These studies emphasize the remarkable similarities between the contractile protein complexes in myxomycete plasmodia and animal muscle.

Some of the above discoveries, as well as their own investigations, have led Jahn and co-workers to propose that ATP-induced contraction of the gel walls of the plasmodial veins forces the flow of sol protoplasm through them. This is preceded by an increasing ATP level that brings about gelation of the protoplasm, accompanied by increasing viscosity of actomyosin, and finally the contraction of this protein, after which the contracted protein molecule goes into solution. It is the gel nearest the sol–gel interface of the vein that first goes into solution, followed by adjacent molecules and fibers of the gel layer. This weakens the vein and subjects it to stretching by the hydrostatic pressure that continually exists within the plasmodium. The new sol formed during contraction flows to an expanding area of the plasmodium and thickens to form gel, which soon repeats the contraction process and reverses the direction of flow. The theory is an attractive one and has received considerable support.

Obviously, in order to explain streaming throughout a large plasmodium, it would be helpful to have evidence of wavelike contractions proceeding in the direction of flow. Although several investigators reported no evidence of such contractions, Rhea (1966b), using a combination of interesting techniques that included time-lapse microcinematography, concluded that peristaltic waves of contraction do occur along the plasmodial strands. Nakajima and Hatano (1962) discovered the presence of acetylcholinesterase in the plasmodium, which led them to suggest that this substance, known to play an important role in nerve impulse transmission, might function in the plasmodium as part of a primitive neuromotor system in protoplasmic movement.

A theory of protoplasmic streaming based on internal water distribution has been presented by Park and Robinson (1967). They find a correlation between vacuolar expansion and contraction and protoplasmic movement in the anterior lobes of the plasmodium and suggest that contraction of vacuoles in this area results in an influx of protoplasm, whereas vacuolar expansion results in an efflux. They further propose that the viscosity of cytoplasm forming the gel layer increases as water passes from the ground plasm into the vacuolar system and that its viscosity decreases with the reverse movement of water. These changes are thought to produce the motivating forces for protoplasmic streaming and regulation of polarity in the migrating plasmodium. The authors of this concept admit to the possible coexistence of their proposal with the contraction–hydraulic theory. The unexplained presence in plasmodia of clear vacuoles that are neither contractile nor food vacuoles requires that this concept be considered in future investigations of protoplasmic flow.

Although considerable attention has been given to the morphogenetic role of cyclic AMP in cellular slime molds, little is known about its possible function in myxomycetes. Murray *et al.* (1971) have found an extensively released cyclic AMP phosphodiesterase in the liquid growth medium of *P. polycephalum* plasmodia and suggest that it may have the protective function in nature of destroying compounds (e.g., cyclic AMP) produced by associated organisms that would otherwise disrupt the normal regulatory mechanism of the myxomycete. This idea has received support from studies by Kuehn (1971), who reports evidence of a more intricate regulatory role for cyclic AMP, by way of its effects on protein kinases, than has previously been proposed for these organisms.

4. Plasmodial Encystment

Plasmodial encystment has been studied most intensively in *P. polycephalum*, in which major portions of the plasmodium develop into relatively firm, irregular, yellowish masses called "sclerotia." Sclerotial development is thought to be influenced by such factors as decreased moisture, lower temperatures, exhaustion of nutrients, and aging. In nature, sclerotized plasmodia may be found within rotting logs during the winter. In the laboratory, sclerotial development in *P. polycephalum* is readily induced by subjecting plasmodia to slow drying. However, the fact that neither drying nor low temperature is necessary for sclerotization is illustrated by the ability of plasmodia of this species to form numerous sclerotia in liquid shake cultures at room temperature. Sclerotium development can be synchronized by the transfer of microplasmodia to a liquid, nonnutrient salts medium (Daniel and Rusch, 1961).

In phaneroplasmodial species, such as *P. polycephalum*, the sclerotium (Fig. 151) is a mass of encysted, mostly plurinucleate protoplasts of various sizes and shapes sometimes called "spherules." The latter are apparently comparable to the separately encysted protoplasts of similarly varied size and shape produced by aphanoplasmodia of other myxomycetes and by plasmodia of certain protostelids.

During sclerotial development protoplasmic streaming ceases as gelation of the entire protoplast occurs (Jump, 1954). This is accompanied

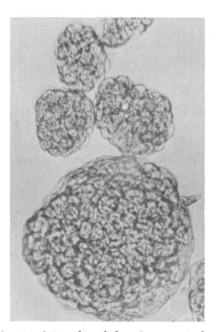

Fig. 151.   Microsclerotia of *P. polycephalum* from axenic liquid culture, showing spherules, ×270. Supplied by D. G. Humm.

by the appearance of Golgi bodies and a copious production of slime similar to that produced by the growing plasmodium (McCormick *et al.*, 1970b). The protoplasmic mass then cleaves into many plurinucleate portions (Stewart and Stewart, 1961; Stiemerling, 1971). The lines of cleavage are determined by numerous vacuoles, reported by Stiemerling to be slime vacuoles and food vacuoles, that become flattened, extended, and fused to each other, thus delimiting a system of walls around the protoplasts. The encystment stage has less water and a higher lipid content than have active plasmodia. The walls of the spherules, as do those of the spores, contain a single polymerized sugar, galactosamine, in addition to a glycoprotein (McCormick *et al.*, 1970b). The presence of this sugar, which appears to be of rare occurrence in nature, is thought to indicate that the myxomycetes are not closely related to either the fungi or the euprotists. During sclerotization there is a rapid turnover of RNA, some classes of RNA decreasing while others increase as differentiation proceeds (Chet, 1973). There is some evidence that carbohydrate metabolism in developing sclerotia is stimulated by cyclic AMP (Lynch and Henney, 1973).

In general, sclerotia retain their viability for a period of 1–3 years.

Those of *Badhamia utricularis* (Bull.) Berk. reportedly remain viable after 7 years of dry storage (Schinz, 1920). The nuclei remain in the unreplicated G_1 phase until advent of the S_1 phase during germination (Hütterman, 1973). When a sclerotium is placed in or on a medium suitable for trophic development, the process of encystment is reversed

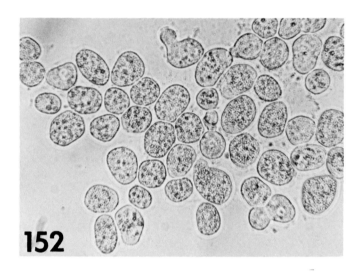

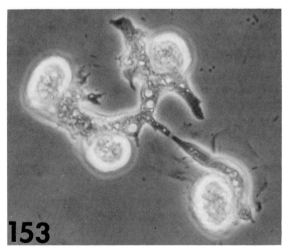

Figs. 152 and 153. Stages in the myxomycete life cycle. Fig. 152. Plasmodial cysts of *Stemonitis* sp., ×250. Fig. 153. Same, germinating; protoplasts fusing, ×250. Figs. 152 and 153 courtesy of K. B. Raper.

as the protoplasts from the numerous germinating spherules rehydrate and merge to form a plasmodium.

Aphanoplasmodial slime molds do not form thick sclerotia. Instead, cysts of variable size and shape are produced directly by the reticulate plasmodium. They thus tend to lie in a single layer on the substrate following the outline of major portions of the plasmodium, at least in laboratory culture (Fig. 152). On germinating, the protoplasts from neighboring cysts fuse together to form a plasmodium (Fig. 153). There is little information on encystment in protoplasmodial forms. A number of protoplasmodial and aphanoplasmodial species occur on bark of trees, a substrate that is subject to rapid wetting and dehydration. Alexopoulos (1964) has pointed out the survival value of the probable ability of these plasmodia in nature to undergo rapid encystment during the approach of dry periods and of the cysts to germinate and rapidly form plasmodia that sporulate before the next dry period.

D. Fruiting

1. Factors Influencing Sporulation

Fruiting of myxomycetes in nature is preceded by the outflow of plasmodia from rotting logs, humus, etc., onto exposed surfaces. Little direct information is available on factors that bring about sporulation under natural conditions, but laboratory studies have shown that sporocarp development is stimulated by starvation, aging, substances in the medium, changes in pH, and, for certain plasmodia, exposure to light. Gray (1938) studied yellow-pigmented plasmodia of four different species, including *P. polycephalum,* and nonpigmented plasmodia of a larger number of species and found that the former require exposure to light before fruiting can be initiated, whereas the nonpigmented plasmodia fruit equally well in light or darkness. In *P. polycephalum* it has also been determined that shorter wavelengths (436–546 nm) of visible light are most effective. For this species Daniel and Rusch (1962) recommend the following conditions for inducing fruiting of plasmodia grown in axenic liquid culture: (1) optimal growth age, coinciding with exhaustion of nutrients; (2) 4 days of dark incubation at 21.5°C on a medium containing only inorganic salts and niacin; (3) exposure to light in the wavelength range of 350–500 nm for 2 hours, followed by a return to darkness. When these conditions are fulfilled, sporulation occurs in 12–16 hours.

Ling (1968) found that *Didymium iridis* fruited best when the brownish plasmodia were grown in the dark for 4 days or more and then exposed to light. Such plasmodia proved highly sensitive to light, showing some fruiting after a 5-second exposure and nearly 100% fruiting after a 1-minute exposure. Straub (1954) made the interesting observation that the period of illumination could be shortened by feeding the test plasmodia on portions of others that had been previously illuminated, indicating the presence of a photosynthetically synthesized substance that helped trigger the fruiting process.

Although a number of investigators have considered the yellow pigments of myxomycete plasmodia as photoreceptors associated with sporulation, there have been conflicting reports on their chemical nature. They have been variously reported to be pteridines, polyenes, flavones, peptide-like pigments, and phenolic compounds. Lieth (1954), using electrophoresis, separated four pigments from the reddish-brown plasmodia of *D. nigripes* but his study of their absorption spectra led him to question whether they are directly responsible for the effective light absorption. The fractionation experiments by Rakoczy (1973) on *P. nudum* Macbr. yielded 12 different fractions, all having the characteristics of pteridines. The action spectrum for fruiting fell within the wavelengths of 330–540 nm and 630–710 nm, with 330 nm being the most effective. Both Lieth (1956) and Rakoczy found that green light inhibits fruiting. It has been proposed that one of the effects of light is to inhibit mitochondrial respiration, thereby reorienting the energy of metabolism and setting the plasmodium on an irreversible course toward sporulation (Daniel, 1966; Cummins and Rusch, 1968). For more detailed discussions of investigations on pigmentation and the effects of light on sporulation, the reader is referred to the publications by Gray and Alexopoulos (1968) and by Rakoczy (1973).

Exhaustion of nutrients in the medium has long been known to contribute toward the initiation of sporulation. The pH of the medium is also of some importance. Gray (1939) found a medium of comparatively high acidity (pH 3) and a temperature of 25°C to be most favorable for fruiting in *Physarum polycephalum*, whereas Collins and Tang (1973) report 100% fruiting in this species at pH 5.

There is also evidence that sporulation can be stimulated by inactivating sulfhydryl (—SH) groups in the plasmodium with compounds, such as *p*-chloromercuribenzoate, that react with them (Ward and Havir, 1957). Plasmodia treated in this manner fruit in about half the time required by untreated control plasmodia.

For some time it has been known that plasmodia on the pathway to sporulation reach a point at which they cannot be induced to revert to the trophic stage. There is now evidence to show that plasmodia of *P. polycephalum*, about 2 hours after the period of illumination that leads to fruiting, contain all the messenger RNA's required for sporulation (Rusch, 1968). Approximately 3 hours after the illumination period these plasmodia are unable to revert to the trophic stage when transferred to fresh growth medium.

2. Sporocarp Development

The simplest form of fruiting in slime molds is found in such protoplasmodial genera as *Echinostelium* and *Licea,* which also have minute sporocarps. Because *Echinostelium* appears to be the genus through which the myxomycetes have evolved from protostelids, it deserves special attention. The smallest known myxomycete, *E. lunatum* (Olive and Stoianovitch, 1971c), has minute amoeba-like protoplasmodia with only two to six nuclei (Figs. 120 and 121). Sporulation begins when a protoplasmodium assumes a hemispherical shape on the substrate, secretes a covering sheath, and produces a narrow and apparently tubular stalk as in protostelids, the protoplast rising on the elongating stalk tip. There is some evidence that a steliogen is present, as in protostelids. As the protoplast leaves the surface of the substrate, much of the sheath that covers it collapses onto the substrate, where it serves as a supporting disc for the sporocarp. This disc is continuous with the sheath surrounding the stalk and sporangial protoplast. When stalk elongation ceases, a mitotic nuclear division occurs in the sporangium and four to eight (rarely more) spores are formed by cleavage of the protoplast. The thin sheath surrounding the spore mass collapses around it and assumes the outline of the spores. Just before the mitotic division in the sporangium, a hollow crescent-shaped structure, believed to be a columella, develops within the protoplast and becomes attached to the tip of the stalk (Fig. 117). Small areolae (hila) are present on the walls of mature spores where they come into contact with each other.

Many features of the above-described spore development are strikingly similar to sporulation in the flagellate protostelid *Cavostelium bisporum,* in which only 2-spored fruiting bodies are produced. The comparison leads us to conclude that the discoid base of the sporocarp, derived from the sheath, may be considered a primitive hypothallus, whereas that part of the sheath around the spores may be viewed as the primitive peridium. A similar primordial type of peridium is characteristic of many members of the Stemonitidae. The more conspicuous and more highly developed hypothallus and peridium of most myxomycetes have probably evolved in conjunction with the tendency for additional materials to be deposited in or on them. Probably in like manner, the stalks of most myxomycetes have become more complex, with waste materials and sporoid bodies formed by the protoplast often being present in them. This is not true of many of the Stemonitida, which have simpler stalks composed of one or more hollow tubes. The above discussion suggests that this order is closely related to the Echinosteliida.

Sporangial development in myxomycetes with reticulate plasmodia begins with the concentration of protoplasm in a number of multinucleate mounds, each of which produces a single sporangium. A careful study of stages in the conversion of plasmodia into sporangia has been made by Lucas *et al.* (1968) in *Didymium nigripes.* Fruiting was induced by transferring plasmodia from the growth medium to nonnutrient agar, after which the following developmental stages were distinguished: (1) advancing front stage or migrating phase, lasting about 16–18 hours; (2) accumulation stage, at which migration ceases and the plasmodial fronts become broader and thicker as the veins empty their contents into them; (3) mounding stage, when the plasmodium pushes up into ridges that become subdivided into presporangial mounds of firmer consistency than the plasmodium; (4) rising stage, during which the pulsating protoplast rises on the developing stalk, the pulsations representing a continuation of the shuttle movement of protoplasm found in the plasmodium; (5) white fruit stage, in which the protoplasm in the sporangial protoplast continues to pulsate for about an hour after completion of the stalk; and (6) pigment synthesis stage, when the sporangium changes from pink to brown and finally black in color as melanin-like pigments are deposited in the spore walls. Stages 1–6 occur in a period of about 32–34 hours, with near synchronous development of all sporangia of the same plasmodium taking place. A number of these developmental stages have been beautifully illustrated by Cohen (1965) for *D. iridis* (Fig. 154). The irreversible commitment to fruiting mentioned earlier occurs during the accumulation stage, after which transfer of the plasmodium to nutrient agar with food organisms fails to cause its reversion to the trophic stage.

Prior to fruiting in *Stemonitis herbatica* Peck, the aphanoplasmodia are reported to develop into ivory white to light yellow coralloid masses that may migrate over the substrate for as long as several days before settling down and producing clusters of sporangia (Indira and Kalyanasundaram, 1963; Indira, 1971). In *Stemonitis* the stalk, which is formed within the rising protoplast, has been described as a single hollow tube, and in the related genera *Comatricha* and *Lamproderma,* as fascicles of narrow tubes (Ross, 1957b).

Descriptions of the fruiting process in *Physarum polycephalum* (e.g., Guttes *et al.,* 1961) indicate plasmodial mounding to form sporangial primordia and a mode of stalk and sporangial development similar

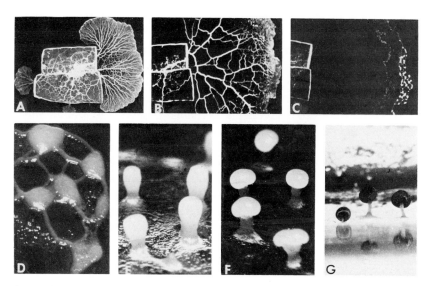

Fig. 154. *Didymium nigripes;* stages in morphogenesis: (A) phaneroplasmodium moving out over agar surface; (B) protoplasm concentrating in outer margin of plasmodium; (C) presporangial masses forming; (D) enlargement of the latter; (E–G) stages in sporangial development. Courtesy of A. L. Cohen (1965).

to that in *D. nigripes.* Most of the pigment granules in the plasmodium are left in the stalk along with other deposits and remnants of the protoplast. The sporangia are somewhat discoid to gyrose and often in a compound cluster. The ashy-colored peridium is covered with squamules of calcium carbonate.

Aethalia and plasmodiocarps are typically sessile fruiting bodies, generally larger than sporangia and, unlike sporangia, they show considerable variation in size within the same species. As noted earlier, aethalia develop from all or a major portion of the plasmodium and are mostly hemispherical or cushion-shaped, whereas plasmodiocarps characteristically develop from major parts of the plasmodial veins and are usually branched or reticulate in form. There appear to be no significant differences between these sporocarps and sporangia in the development of the sporogenous portion of the fruiting body.

During sporocarp development the protoplasm loses a great deal of water. In *P. polycephalum,* for example, the water content falls from about 80% in the plasmodium to about 20% in the spores (Daniel, 1966). Much of the water is excreted onto the surface of the sporocarp, and in many slime molds, as the water evaporates, its salts are deposited on the peridium. In *Fuligo* this results in the encrustation of the aethalium with thick deposits of salts as well as pigment granules. In *Didymium* a network of tubules is reported to develop within the sporangial protoplast and to act as a system of channels through which the calcium carbonate in solution is removed and deposited at the surface of the peridium (Welden, 1955). The carbonate may occur in either amorphous or crystalline form in the different genera. In species in which the calcium carbonate is not completely eliminated from the sporangium, it may be deposited in the capillitium or columella. However, in some myxomycetes the presence or absence of readily observable amounts of calcium carbonate is highly variable under different growth conditions, which can make keys emphasizing this feature frustrating.

In quite a number of myxomycetes no capillitium is formed within the sporocarp, whereas in others a well-developed elaborate one is present. In *Stemonitis,* according to Ross (1957b), hollow capillitial threads originate from the columella, which is a lengthy continuation of the stalk into the sporangium, and from the periphery of the sporangium where it forms a surface network. When mature, the entire capillitium is united into a single system connected to the central columella (Fig. 137). Both Ross (1957b) and Goodwin (1961), who have studied species of the related genus *Comatricha,* conclude that the capillitium is formed by intraprotoplasmic secretion and not by special vacuoles and invaginations as claimed for a number of myxomycetes; whereas Mims (1973) finds that the hollow capillitial threads of *Stemonitis virginiensis* Rex are formed within an anastomosing system of tubular vacuoles. The peridium, as in *Echinostelium,* is no more than a single unit membrane. Mims' excellent photographs illustrate the major stages in conversion of the aphanoplasmodium of *Stemonitis* into sporangia (Fig. 155). The dark stalk is apparent within the culminating sporangial protoplast.

In *Badhamia gracilis* (Macbr.) Macbr. of the Physarida, a system of tubular invaginations reportedly extends from the periphery of the sporangial protoplast inward. This is added to by vacuoles within the protoplast, following which the capillitium develops within the resulting network of channels (Howard, 1931; Welden, 1955). *Physarum polycephalum* has a similar precapillitial network of channels (Howard, 1931), although it is uncertain whether or not peripheral invaginations contribute to it.

Many years ago Harper and Dodge (1914) described the capillitium of *Trichia* and *Hemitrichia* as originating within a system of tubules derived from the fusion of vacuoles in the protoplast. The hollow capillitial threads arise by deposition of materials around the periphery of these tubules. Further depositions on the surfaces of the threads produce the characteristic spiral thickenings (Fig. 136). Ultrastructural details of

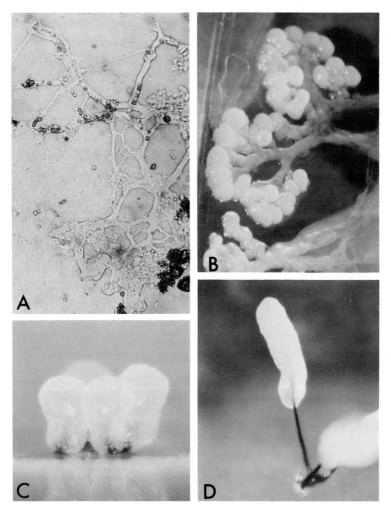

Fig. 155. *Stemonitis virginiensis:* (A) aphanoplasmodium, ×200; (B) development of presporangial masses, ×8; (C) beginning of stalk formation, ×12; (D) later stage in sporangial development, ×10. From Mims (1973).

capillitial characteristics and mode of development are described in Section VII.

3. Spore Formation

The sporangial protoplast of myxomycetes cleaves progressively into small uninucleate segments that secrete walls around themselves and become spores. It has generally been concluded that at some point in

the sporulation process meiosis occurs. Without going into all the conflicting ideas that have been published on the nature and position of meiosis
in the life cycle of myxomycetes, some concept of this diversity of opinion
may be obtained from a few selected publications.

Wilson and Cadman (1928) reported that meiosis in *Reticularia
lycoperdon* occurred in two successive nuclear divisions prior to spore
delimitation, a conclusion later extended by Wilson and Ross (1955)
to several other species. Later, Ross (1961) concluded that the second
meiotic division occurred in some species while spore cleavage was in
progress. Mims (1973) has found a synchronous mitosis at the time
of cleavage in *Stemonitis*, during which "protospores" are formed in
which the nuclear division is completed and each protospore divides
into two uninucleate spore cells, but the division was concluded to be
nonmeiotic. Jahn (1933) could find evidence of only one nuclear division
prior to spore formation in young sporangia of *Badhamia utricularis*.
Nevertheless, he concluded that it was meiotic in nature and that the
second meiotic division occurred at spore germination. Von Stosch
(1935) also observed a single synchronous precleavage nuclear division
in *Didymium nigripes* but concluded that it was an equational mitotic
division. In the maturing spores, however, he found two successive nuclear divisions that he thought were meiotic. One nucleus was reported
to degenerate after each division, thus leaving the spore with a single
haploid nucleus ($n = 24$–25).

In recent years several investigators, using electron microscopy, have
confirmed von Stosch's concept as to where meiosis occurs in the life
cycle. Aldrich (1967), in a study of three species of *Physarum* (including
P. polycephalum) was the first to demonstrate the presence of synaptonemal complexes—indicators of meiotic chromosomal pairing—in the maturing spores. The frequency of these complexes, in addition to Aldrich's
confirmation of von Stosch's observation that two successive nuclear
divisions occur in the maturing spores, leaves little doubt that meiosis
occurs here. Three of the four resultant nuclei were thought by Aldrich
to degenerate, leaving a single haploid nucleus in the spore. These observations were later extended by Aldrich (1970) and Aldrich and Mims
(1970) to members of all other orders of Myxogastria except the Echinosteliida. Finally, Haskins *et al.* (1971) discovered synaptonemal complexes
in spores of *Echinostelium minutum*, thus embracing the fifth and final
order in this group. This leads to the conclusion that meiosis in myxomycetes typically occurs in the maturing spores within the sporocarp. Additional confirmation is found in the absence of synaptonemal complexes
in nuclei prior to cleavage (Aldrich and Mims, 1970).

It is of particular interest to note that, of the above-mentioned myxomycetes found to have synaptonemal complexes in their maturing spores, *E. minutum* and *Physarum pusillum* complete their life cycles in single-spore culture. Therefore, these species appear to be homothallic rather than apomictic, although they may be apogamic (syngamy not having been demonstrated).

In a comparative statistical estimation of the number of nuclei present in young sporangia and the number of spores present in mature sporangia, LeStourgeon *et al.* (1971) found that the spore number was approximately twice the number of nuclei in young sporangia, indicating that there is only a single precleavage mitosis in the sporangium. The claims of several earlier investigators that precleavage meiosis could be deduced from the observed presence of "typical meiotic prophase configurations," chromosomal bridges, etc., are probably indicative of the unreliability of these criteria when applied to such organisms as myxomycetes with numerous small chromosomes. A similar discrepancy appears to occur in a spectrophotometric analysis of *P. flavicomum* by Therrien (1966), who reported that nuclei at the time of spore cleavage contain a haploid amount of DNA.

VII. ULTRASTRUCTURE

A. Haploid Trophic Stage

On germinating, the spore, which typically has a single haploid nucleus, produces from one to several uninucleate amoeboid or flagellate cells. The release of more than one cell from the spore is generally thought to be preceded by mitotic division and cytokinesis. In an ultrastructural study of spore germination in *Arcyria cinerea*, Mims (1971) found that most spores were uninucleate, although some were plurinucleate, before the onset of germination. Early signs of germination were observable in the disappearance of presumed polysaccharide material and lipid bodies, a less dense appearance of the cytoplasm, and more distinct mitochondria and endoplasmic reticulum (ER). In addition, large vacuoles, a Golgi apparatus, and a pair of centrioles appeared in the cytoplasm. The single uninucleate cell emerged from the spore through a break in a bulge of the wall that appeared at one end of the spore.

The slime mold myxamoeba becomes flagellate when the centrioles function as kinetosomes and give rise to flagella (Schuster, 1965a; Mims, 1971), usually a long anteriorly oriented one and a short laterally appressed one. Neither has mastigonemes. Schuster has described the transformation of the centrioles in *Didymium iridis* and *Physarum cinereum* as involving movement of the centrosphere, including the Golgi body and centrioles, to the cell periphery, where the latter bodies (now kinetosomes) elaborate the flagella. He also found centrioles in microcysts, indicating their potentiality to form flagellate cells during germination. In *P. flavicomum* Berk., Aldrich (1968) observed that a flagellar vesicle develops at the distal end of the kinetosome and, as the flagellar bud elongates into it, forms the flagellar sheath. Just before they emerge from the cell, both flagellar buds can be found within a common vesicle. The usual cell organelles, including uninucleolate nucleus, Golgi apparatus, mitochondria, rough ER, food vacuoles, and contractile vacuole, are found within myxamoebae and flagellate cells.

The flagellar apparatus (Fig. 156) has been studied in considerable detail in *P. flavicomum* (Aldrich, 1968). Each flagellum shows the conventional 9 + 2 pattern of microtubules. The anterior end of the cell is held in a more rigid condition than the posterior end by means of microtubules that radiate from the kinetosomes back into the cytoplasm. There is a conelike arrangement of two arrays that arise from the kinetosome of the long flagellum, the microtubules of the outer array ending at about the level of the posterior end of the nucleus. A sparser, single array of microtubules arises from the base of the short flagellum and runs posteriorly just beneath the cell membrane. The Golgi apparatus

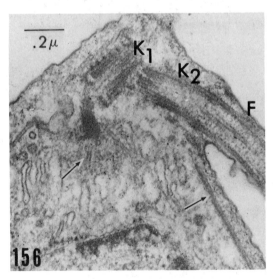

Fig. 156. *Physarum flavicomum;* electron micrograph of flagellar apparatus, showing kinetosomes (K₁, K₂), short reflexed flagellum (F), and microtubules emanating from kinetosomes (arrows). From Aldrich (1968).

is included within the flagellar cone, which appears to be more or less equivalent to what has previously been called the centrosphere.

Nuclear division in the haplophase is extranuclear (astral) and is characterized by breakdown of the nuclear envelope in late prophase, followed by the appearance of a distinct spindle with polar centrioles (Aldrich, 1969). If flagella are present, they are resorbed prior to the division. The cell constricts into two daughter cells and a new nuclear envelope is reorganized around each daughter nucleus. The generation time of myxamoebae grown in axenic culture has been determined as 14.5 hours for *Didymium nigripes* and nearly 9 hours for *Physarum cinereum*.

In an ultrastructural study of *P. flavicomum* myxamoebae in the process of digesting cells of *Escherichia coli*, Kazama and Aldrich (1972) found acid phosphatase to be localized in food vacuoles, Golgi apparatus, and small membrane-bound vesicles. They also suspected the presence of this enzyme in smooth ER. They note that in a number of protozoans acid phosphatase is known to reach the food vacuoles by way of vesicles derived from ER or Golgi bodies, or from acid phosphatase-rich ergastoplasm that concentrates around the food vacuoles.

B. Plasmodium

The conventional eukaryotic organelles have been found in thin sections of myxomycete plasmodia, including many uninucleolate nuclei, Golgi bodies, mitochondria with tubular cristae, smooth and rough ER, vesicles and vacuoles of various types, polyribosomes, microfibrils, and microtubules (McManus and Ruch, 1965; McManus and Roth, 1965). There is some evidence that one type of vesicle is blebbed off by the nuclear envelope. A layer of cytoplasm just beneath the plasma membrane appears to be devoid of cell organelles (Charvat *et al.*, 1973b).

Electron microscopy has confirmed earlier observations that the nuclear envelope persists during mitosis in the plasmodium and that an intranuclear spindle lacking polar centrioles is produced (Guttes *et al.*, 1968). The nucleolus breaks up into fragments that remain identifiable until late anaphase, and each new nucleolus is derived from the fusion of several pronucleoli dispersed among various telophase chromosomes. An amorphous, granular or fibrous spindle primordium that appears within the nucleus during prophase is thought to divide, and short microtubules radiating from the two resultant bodies elongate to form the intranuclear spindle (Sakai and Shigenaga, 1972). Kinetochores occur at points of attachment of microtubules to chromosomes. As the daughter chromosomes move to opposite poles during anaphase, a localized breakdown of the nuclear envelope can be seen at the spindle poles as well as around the middle area where the daughter nuclei separate. The missing parts of the nuclear envelope are replaced during telophase.

As in other organisms, there is evidence that the mitochondria of myxomycetes contain DNA. The plasmodial mitochondria have been found to incorporate tritiated thymidine to a conspicuous degree (Guttes and Guttes, 1964). In addition, the mitochondrion is known to possess an elongate fibrous core (Niklowitz, 1957) that appears to divide just before mitochondrial division (Schuster, 1965b; Guttes et al., 1966; Charvat et al., 1973a). The fibrils of the core resemble the DNA fibrils found in mitochondria and chloroplasts of higher organisms. The core is digested by deoxyribonuclease but not by ribonuclease. The mitochondrial DNA has a molecular weight of 2×10^7 to 3×10^7 and is therefore a much larger molecule than that of vertebrate mitochondria (Evans and Suskind, 1971). A most remarkable occurrence has been reported in the plasmodia of an isolate of *Didymium squamulosum* (Alb. & Schw.) Fr. (Duval, 1970). The mitochondria become invaded by a rickettsia-like microorganism, and after their degeneration the parasite is released into the cytoplasm. The parasite is also present in the spores. It causes degeneration of the plasmodium but can be eliminated with chloramphenicol.

Wide interest in the mechanism of amoeboid movement has naturally focused attention on the microfibrils of the plasmodium. Wohlfarth-Bottermann (1964) found many of them occurring in parallel arrangement within bundles that closely resembled smooth muscle fibers. He and Charvat et al. also noted that some of the microfibrils joined the plasma membrane in a manner indicating their possible involvement in protoplasmic streaming. McManus and Roth (1965) observed similar microfibrils, often in conspicuous bundles but not inserted on the plasma membrane, in protoplasmodia of *Clastoderma debaryanum* Blytt and phaneroplasmodia of *Fuligo septica* but not in the aphanoplasmodia of *Stemonitis fusca* Roth. Rhea (1966a) reported microfibrils in *Physarum polycephalum* positioned in circular and longitudinal arrays in the cortical cytoplasm in such a way that, if contractile as suspected, they would be able to bring about changes in diameter and length of plasmodial regions. These changes, in turn, could result in streaming and motility. Fibrils were also found in contact with the plasmalemma in these areas. They measured 50–70 Å in diameter and were similar to the ATP-sensitive actomyosin filaments extracted from plasmodia (Ts'o et al., 1957). Usui (1971), who found loose networks and compact bundles of microfibrils associated with pseudopodia, also concluded that they generated the force of streaming.

From plasmodia of *P. polycephalum* Hatano and Oosawa (1964) isolated an actin-like protein that formed an actomyosin-like complex with rabbit muscle myosin. Nachmias and Ingram (1970) obtained myosin-enriched actomyosin by removing some of the actin and observed series of arrowhead structures similar to those of vertebrate striated muscle actomyosin. When myosin was separated from actomyosin and then precipitated, the precipitate was found to contain aggregates of short bipolar filaments similar to those observed in sectioned material. Alléra and Wohlfarth-Bottermann (1972) found bundles in which the ultimate observable substructure was interpreted to be filaments of myosin 20–25 Å wide, occasionally connected with actin by crossbridges. They proposed that transformation of stationary ectoplasm into streaming endoplasm is accompanied by disaggregation by myosin aggregates into monomeric and oligomeric units for transportation. The reverse process would produce contraction. The cyclic nature of these changes anteriorly and posteriorly in the plasmodium would presumably result in the shuttle movement of protoplasm. Further evidence for this viewpoint was found in the observation that "contraction islets" containing abundant thick fibrillar bundles appeared following fixation of plasmodia in a paraformaldehyde solution containing ATP.

Ultrastructural events accompanying sclerotization of plasmodia have also been the subject of recent investigation (Goodman and Rusch, 1970). During the process glycogen granules decrease in number and Golgi complexes thought to be associated with elaboration of polysaccharides appear. Cytokinesis is accomplished by the fusion of vesicles. Following cleavage, smooth ER changes to rough ER. Experiments with cycloheximide, which inhibits protein synthesis, and actinomycin, which blocks RNA synthesis, indicate that the proteins necessary for cleavage are synthesized about 20 hours after sclerotization is stimulated by starvation but that new RNA synthesis following such stimulation is not required for sclerotization.

C. Sporogenesis

Both light and electron microscopy indicate that a single synchronous, nonmeiotic, precleavage nuclear division occurs in the sporocarp, during which the nucleolus disappears. Studies on *Arcyria cinerea* (Mims, 1972a) demonstrate that the division is intranuclear, without centrioles, and essentially synchronous throughout the sporangium, thus resembling mitosis in the plasmodium. Spore cleavage follows this division. In *Stemonitis* Mims (1973) found a somewhat different sequence of events. Mitosis follows cleavage of the sporangial protoplast into uninucleate prespore cells, after which the resultant binucleate cells divide to form two spores each. The mitotic division is unaccompanied by centrioles, and the nuclear membrane breaks down only at the poles (Fig. 157). In *Didymium nigripes* some of the nuclei in the developing sporangium become incorporated into autophagic vacuoles where they disintegrate (Schuster, 1969). These vacuoles, which include other organelles such as mitochondria, become incorporated into the spores, where they are

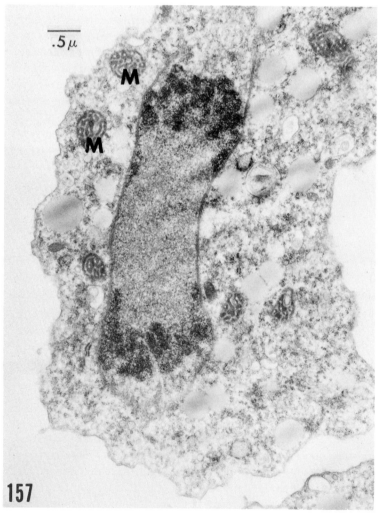

Fig. 157. *Stemonitis virginiensis;* electron micrograph of telophase of mitosis in protospore. Note polar breaks in nuclear envelope; mitochondria (M). From Mims (1973).

thought to be a source of nucleic acid and protein at the time of germination. Cronshaw and Charvat (1973), who found a high degree of β-glycerophosphatase activity in both autophagic and food vacuoles, consider them comparable to lysosomes of higher organisms.

Spore cleavage has been studied in sporangia of *D. nigripes* and *Physarum flavicomum* by Schuster (1964a) and Aldrich (1966), respectively. In a reinvestigation of *D. nigripes,* Aldrich (1974) found that cleavage begins with invagination of the plasmalemma, the cleavage furrows thereafter being extended by the addition of Golgi-derived vesicles to their advancing ends. Spines are laid down around the developing spores, and these occur opposite each other in adjacent spores. A continuous spore wall is deposited beneath the spines. The spore wall of *Physarum* is comprised of a galactosamine polymer, a glycoprotein, and melanin, thus resembling in composition the spherule wall (McCormick *et al.,* 1970b). This contrasts with earlier reports of cellulose (Goodwin, 1961) and of cellulose and chitin (Schuster, 1964a) in myxomycete spore walls.

The discovery of synaptonemal complexes in the maturing spores of one or more members of all orders of the Myxogastria (Aldrich and Mims, 1970; Haskins *et al.,* 1971) is strong evidence that the developing spore is the normal site of meiosis in myxomycetes. A previous ultrastructural investigation demonstrating synaptonemal complexes (Carroll and Dykstra, 1966), presumably during prophase of precleavage mitosis, now seems to have been in error with regard to time of occurrence (Aldrich, 1970). The synaptonemal complexes are of characteristic appearance, consisting of a central component and two lateral elements (Figs. 158 and 159). Polycomplexes have been found within the nucleolus of the spore nucleus in both *Stemonitis herbatica* (Aldrich and Mims, 1970) and *Echinostelium minutum* (Haskins *et al.,* 1971). Synaptonemal complexes have also been found in some older spores from mature sporangia in *Stemonitis virginiensis* (Mims and Rogers, 1973), which may be indicative of interrupted meiosis in these particular spores. If such spores are viable, this behavior must be taken into account in genetic analysis.

Meiotic prophase is generally recognizable about 18–24 hours after cleavage and is followed by two successive nuclear divisions. Both meiotic divisions have been described as intranuclear and unaccompanied by centrioles (Aldrich, 1967). The nuclear envelope becomes discontinuous at the spindle poles (Fig. 160) and is reconstituted in these areas during telophase. Three of the four resultant nuclei degenerate, leaving a single haploid nucleus in the spore, at which time Golgi apparatus and a pair of centrioles appear in each spore (Aldrich, 1967; Mims, 1972b). Haskins *et al.* (1971), using both light and electron mi-

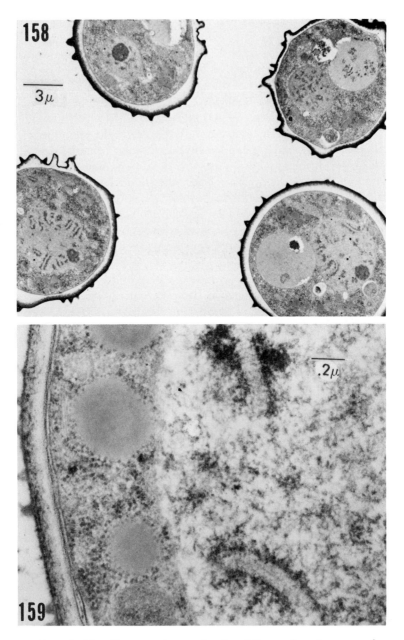

Figs. 158 and 159.   Electron micrographs of meiotic prophase in spores of myxo-mycetes. Fig. 158. Spores of *Physarum pusillum* showing numerous synaptonemal complexes. Fig. 159. Detail of synaptonemal complexes in *Arcyria incarnata*. From Aldrich and Mims (1970).

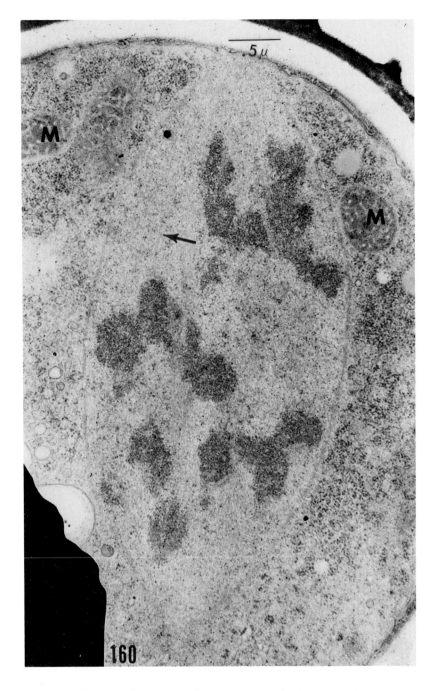

Fig. 160. *Physarum flavicomum;* electron micrograph of spore nucleus at meiotic anaphase. Note numerous spindle microtubules (arrow), chromosomes, and open ends of nuclear envelope; also mitochondria (M). From Aldrich (1967).

croscopy, confirmed the intranuclear nature of the first meiotic division in the spores of *Echinostelium minutum*, but their light microscope observations on the second division indicated a breakdown of the nuclear envelope. Also, they reported that one nucleus degenerates after each division, as claimed earlier by von Stosch (1935) for *Didymium iridis*.

During all stages of sporocarp development, including spore formation, in *Perichaena vermicularis*, a high degree of acid phosphatase activity is reportedly localized in single membrane-bound vacuoles (autophagic vacuoles) containing cytoplasmic organelles in various stages of digestion, as well as in food vacuoles containing cells of *Escherichia coli* in plasmodia of the same species (Charvat *et al.*, 1973a,b). The reaction product was also found in the walls of mature spores. During sporocarp development, an autolytic outer layer is very early apparent. In this region are found autophagic vacuoles, degenerating nuclei, and later extensive channels with degenerating protoplasm in which acid phosphatase activity is high. When autolysis is completed a peridial layer replaces it, with an additional layer being laid down along its inner surface.

The usefulness of scanning electron microscopy in the classification of myxomycetes is well illustrated in the extensive investigations of Schoknecht and Small (1972), who studied the spores and capillitia of 51 species in 21 genera by this method. Spores of all the Myxogastria studied by them possessed some kind of spore ornamentation, even though the markings were sometimes too minute to be resolved by light microscopy. In contrast, spores of the now excluded *Ceratiomyxa fruticulosa* had entirely smooth walls. These authors concluded that basic spore wall morphologies are fairly similar within natural taxonomic groups and that spore characteristics are among the most stable for taxonomic purposes. Ellis *et al.* (1973) studied the capillitia of 11 species of Trichiidae and two of Dianemidae by phase contrast as well as by scanning and electron microscopy and concluded that five major capillitial types were present. These capillitial characteristics, however, appear to have limited applicability to the classification of slime molds. Similar studies of *P. polycephalum* have revealed surface pores in the peridium that are continuous with the lumens of capillitial threads (Kislev and Chet, 1973).

The spore walls of both *Echinostelium lunatum* and the pink-spored form of *E. minutum* appear distinctly punctate under the phase-contrast microscope and spiny under scanning electron microscopy (Fig. 161), the markings being similar to those on spores of the protostelid genus *Cavostelium*. When thin sections of these spores are studied by electron microscopy they show channelized projections passing through the spore

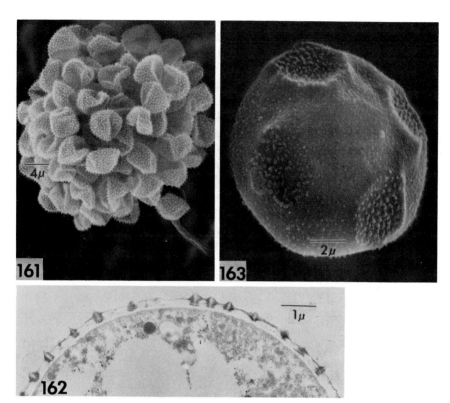

Figs. 161–163. Ultrastructural details of myxomycete spore walls. Fig. 161. *Echinostelium minutum* (pink-spored form); scanning electron micrograph of sporangium showing distinct spore wall protrusions. Micrograph by Michael Dykstra. Fig. 162. *Echinostelium lunatum*, thin section of spore wall showing channelized protrusions. Micrograph by C.-Y. Hung. Fig. 163. *Echinostelium minutum* (white-spored form), scanning EM showing smaller spore wall protrusions (somewhat larger in areas of the areolae). Courtesy of E. F. Haskins.

wall (Fig. 162) very similar to those of *C. bisporum* (Hung and Olive, 1972a). These similarities, among others, are believed to indicate a close phylogenetic relationship between protostelids and the most primitive myxomycetes. The white-spored form of *E. minutum* has much finer spore markings, which are broader in the areolate areas (Fig. 163).

Types of spore germination have been discussed by Mims and Rogers (1973), who studied the ultrastructure of germination in *Stemonitis virginiensis*. The spore wall of that species is comprised of an electron-dense outer layer and a thinner electron-transparent inner layer. However, in the area of germ pore formation the inner wall is thicker and the outer thinner than elsewhere. At germination the outer wall breaks and the inner, more elastic component is pushed out and eventually ruptured by the emerging protoplast.

VIII. GENETICS

Genetic studies of myxomycetes have dealt in large part with types of compatibility (or incompatibility) systems, one controlling gametic fusions and the other mediating the coalescence of plasmodia. There have also been some investigations of factors involving plasmodial pigmentation and drug resistance. Genetic techniques for *Physarum polycephalum* have been reviewed by Dee (1973). Descriptive methods for growing and crossing cultures of *Didymium iridis,* with special emphasis on classroom use in demonstrating the multiple allele mating system and the plasmodial incompatibility mechanism in slime molds, have been published by Collins (1969a).

A. Sexual Compatibility

Skupienski (1927, 1928) was probably the first to demonstrate heterothallism in myxomycetes. He observed that the one or two cells produced by a single germinating spore of *Didymium difforme* (Pers.) Gray were of one compatibility type, and that when a clone from one spore was crossed with a compatible clone of another, sporulating plasmodia developed. Collins (1963) and Collins and Ling (1964) were the first to show that the mating-type system of a myxomycete is controlled by multiple alleles at a single locus. They identified six different mating-type alleles among three different isolates of *Didymium iridis.*

As in many basidiomycetous fungi (Raper, 1966), a gamete carrying one allele (e.g., A^1) is compatible with a gamete carrying another allele (A^2, A^3, A^4, etc.) and is incompatible only with a gamete having the same allele. This, of course, favors outcrossing while limiting inbreeding.

Multiple allelism at the mating-type locus has now been confirmed for a number of other heterothallic species, including *Physarum polycephalum* (Dee, 1966) and *P. pusillum* (Collins *et al.,* 1964), as well as *P. flavicomum, P. rigidum* (G. Lister) G. Lister, and *Fuligo septica* (M. Henney and Henney, 1968). It is likely to prove the characteristic type of heterothallism in this subclass.

Earlier reports as well as those of recent investigators (Mukherjee and Zabka, 1964; Collins, 1961, 1965; Dee, 1966; Collins and Ling, 1968) have noted repeated instances of plasmodial and sporocarp development in single-spore clones of normally heterothallic species. Some of the explanations for this behavior have been as follows: (1) persistence of more than one haploid nucleus following meiosis and emergence from the spore of cells of compatible mating types, (2) presence of a diploid or aneuploid nucleus heterozygous for mating type in the mature spore, (3) mutation at the mating-type locus, and (4) selfing. The likelihood that some spores of *Stemonitis virginiensis* are diploid has already been noted (Mims and Rogers, 1973). Evidence of mutation at the mating-type locus (Collins, 1965) and of selfing (Collins and Ling, 1968) has been obtained in *Didymium iridis*. However, Collins and Ling reported that when selfing occurs sporulation is poor and spore viability very low. Yemma and Therrien (1972) conducted a microspectrophotometric study of nuclear DNA in two of these selfing clones supplied to them by Collins and concluded that nuclei in developing sporangia of selfed clones contained the same amount of DNA as those in sporangia of heterozygous diploids. In addition, mitotic divisions in the plasmodia were infrequent and the chromosomes of the plasmodial nuclei remained mostly in unreplicated condition.

Some of the difficulties inherent in the application of interisolate compatibility to relationships among isolates are illustrated in the study of M. Henney and Henney (1968). The species investigated, *P. flavicomum* and *P. rigidum,* are often difficult to distinguish from each other. A group of isolates that failed to cross with *P. flavicomum* were classified as *P. rigidum.* In similar fashion two other groups were determined to be *P. flavicomum.* Members of the latter two groups, however, proved to be incompatible in intergroup matings in spite of being very similar in appearance. The question remaining is whether these two groups represent two closely related species or a single species in which incompatibility factors independent of the mating-type locus prevent their crossing. Further investigation is needed to resolve this issue.

B. Plasmodial Incompatibility System

Alexopoulos and Zabka (1962), studying *Didymium iridis,* were probably the first to discover that only plasmodia of the same type would coalesce. Coalescence was followed by complete intermingling of the two protoplasts into a single larger plasmodium. This behavior was soon confirmed by Collins (1963, 1966) with regard to the mating type locus in *D. iridis.* Although gametes of a Honduran isolate (A^1A^2) were cross-compatible in all combinations with those of a Panamanian isolate (A^3A^4), the plasmodia of the two failed to coalesce. However, like plasmodia from interisolate crosses coalesced readily if derived from the mating of the same two clones.

It was eventually discovered that a number of different loci, acting primarily as an incompatibility system, were controlling plasmodial coalescence in *D. iridis* (Collins and Clark, 1968; Collins and Ling, 1972; Clark and Collins, 1973a; Ling and Collins, 1970a,b; Ling, 1971). Coalescence occurred when the plasmodia were alike phenotypically with respect to these loci. For example, $CDefGA^1/cdefGA^2$ and $CDefGA^1/CDefGA^2$ plasmodia readily coalesce but not $CDEFGA^1/cdefGA^2$ and $CDefGA^1/CDefGA^2$, because differences at E and F loci in the latter combination make the plasmodia phenotypically different. The incompatibility loci were found to fall into three linkage groups, with one of the markers linked to mating type.

Carlile and Dee (1967) and Poulter and Dee (1968) concluded from their investigations of *Physarum polycephalum* that several alleles at a single locus (f) control plasmodial coalescence in that species, only plasmodia with similar alleles at that locus being able to coalesce. However, subsequent investigations (Collins and Haskins, 1970, 1972; Collins, 1972) have shown that the situation here is like that in *D. iridis* in being under the control of a polygenic system (four loci detected) mediated by simple dominance at each locus.

Results from the laboratories of both Collins and Carlile convincingly demonstrate that these loci function primarily in inhibiting coalescence of dissimilar plasmodia rather than in promoting coalescence of similar ones. Clark and Collins (1973b) believe that this incompatibility system enables a plasmodium to recognize and eliminate "potentially foreign matter"—in this case, the protoplasm of any plasmodium differing in phenotype. In their study of phenotypically dissimilar plasmodia of *D. iridis*, they find that the plasmodia undergo temporary fusions that are quickly terminated by cytotoxic reactions, producing killed areas varying from microscopic to several square centimeters. Each locus allows a characteristic amount of protoplasmic mixing, G being the strongest because it permits the smallest zone and therefore the least amount of killing, whereas E is the weakest, since it allows the greatest amount of intermixing and killing. As the plasmodia differ at more loci, the results are additive in reducing the zone of intermixture. The killing reaction is typically unidirectional, with the cytotoxic effects occurring in the recessive phenotype. In a study of certain diploid heterokaryons that were induced to fruit, it was found that one nuclear type was generally eliminated before sporulation.

Similar lethal reactions between plasmodia have also been reported for *Physarum polycephalum* (Fig. 164), with complete killing of large plasmodia by small ones occurring (Carlile and Dee, 1967; Carlile, 1972). Nuclei of the unfavored plasmodium are eliminated, probably by enclo-

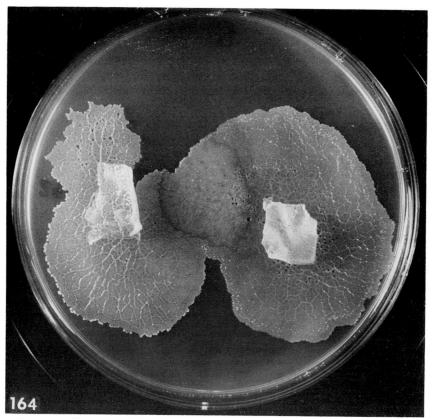

Fig. 164. *Physarum polycephalum;* lethal interaction between two different plasmodial strains. The central killed area results from heterogenic somatic incompatibility manifest following plasmodial fusion. From Carlile (1972).

sure in autophagic vacuoles. Carlile (1973) refers to these postfusion lethal reactions, as well as the failure of plasmodia to fuse, as manifestations of heterogenic somatic incompatibility.

C. Other Genetic Factors

The wild-type color of the *Didymium iridis* plasmodium is commonly brown, the color being produced by a reddish pigment. Two recessive mutant genes have been found that block normal production of this pigment (Collins and Clark, 1966; Collins, 1969b; Collins and Erlebacher, 1969). Plasmodia homozygous for the mutant allele of either gene are cream-colored. Plasmodia containing nuclei heterozygous for both genes are wild type in color, but diploid heterokaryons are noncomplementing and result in cream-colored plasmodia. In other words, gene complementation between these loci is intranuclear but not internuclear.

Mutagenic techniques especially useful in producing drug-resistant mutants in myxamoebae of *Physarum polycephalum* have been described by Haugli and Dove (1972a,b). These methods involve ultraviolet irradiation in conjunction with the use of caffeine, nitrosoguanadine, or ethyl methane sulfate. The eight independent recessive mutants for cycloheximide resistance that were obtained segregated in typical Mendelian fashion.

A homothallic strain of *P. polycephalum* obtained by Wheals (1970) is likely to become a valuable asset to genetic studies, because it permits analysis of the effects of recessive genes in the homozygous state in the plasmodium. An added advantage is found in the ability to cross the homothallic isolate with the heterothallic form. In fact, a mutant locus (*apt*) in the homothallic strain prevents selfing but does not interfere with crossing with similar but nonallelic mutants or with the heterothallic form (Wheals, 1973). A valine-requiring mutant has been studied in these crosses.

Most urgently needed at present are gene mutants that can facilitate analysis of the succession of biochemical events that occur during development in a slime mold. Suitable subjects and techniques are now becoming available in the myxomycetes and will undoubtedly lead to the increased use of these organisms in developmental studies.

Mycetozoa

Acrasea

CHAPTER 5

Acrasea (Acrasid Cellular Slime Molds)

I. INTRODUCTION

The class Acrasea, with the single order Acrasida, is referred to here as the acrasid cellular slime molds, or simply the acrasids. Until recently (L. S. Olive, 1970), they were classified in the same order as the dictyostelid cellular slime molds. However, several unique features in their structure and development strongly favor their separate treatment. E. W. Olive (1902) was well aware of a number of these differences, on the basis of which he recognized two major families, Dictyosteliaceae and Guttulinaceae, in the single order Acrasieae. He considered both amoebae and sorocarps of the former to be more highly differentiated and emphasized distinctions between amoebae with filose pseudopodia in the former and those with lobose ones (or possibly no true pseudopodia) in the latter family. Reasons for the changes in names of major taxa in this class are discussed in Section IV.

Although the acrasids have been the most neglected group of mycetozoans, interest in them has increased in recent years. For the first time, the life cycles of three members of the class have been studied in some detail, and ultrastructural studies have contributed considerably to a better understanding of their structure and development. Consequently, the contrasts between acrasids and dictyostelids (as well as other eumycetozoans) have become even more obvious, resulting in our decision to set up a separate class to accommodate the former group. There is increasing evidence that acrasids and eumycetozoans are not closely related but have had separate phylogenetic origins from ancestral euprotists. The ability to aggregate and form fruiting bodies is therefore believed to have originated independently in the two groups of cellular slime molds, just as it has in the prokaryotic myxobacteria and the unidentified ciliate discussed in Chapter 1.

The life cycles of only three species of acrasids—*Acrasis rosea* L. Olive and Stoian., *Guttulinopsis vulgaris* E. W. Olive, and *Copromyxa arborescens* Nesom and L. Olive—have been studied in any detail in the laboratory. These investigations reveal such striking differences in modes of sorogenesis and in structural details that one is inclined to suspect that the Acrasea themselves have had a polyphyletic origin. In addition, there is now evidence of the direct evolution of at least one acrasid genus from a flagellate ancestor. For the first time in any cellular slime mold, the little-known acrasid *Pocheina rosea* (Cienk.) Loeblich and Tappan has been found to produce flagellate cells (Olive and Stoianovitch, 1974).

The amoeboid cells of *Acrasis rosea,* a species that has been studied in some detail, may be used to exemplify the trophic stage of the class. In water or on an agar surface the amoeba moves by means of rapid, successive hyaloplasmic bulges of the cell. A broad area of hyaloplasm has been observed to appear in the anterior region of the cell and move in a wave posteriolaterally or in two or three such waves, one overlapping the other. A bulge may develop laterally on the cell at any time, often changing the direction of movement. Both anterior and lateral bulges may appear more or less simultaneously on the cell. In water, movement is more rapid, the amoeba more elongate, and unidirectional movement more sustained. One is impressed with the fact that these cells move much more rapidly than those of dictyostelid cellular slime molds. Also, cells in water frequently develop a posterior knoblike uroid region, often with fine filose extensions. The contractile vacuole is generally located within or near the uroid. The filose extensions, which are not found elsewhere on the cell, are adherent to objects and may play a part in the ingestion of food cells. The amoeba typically contains a single nucleus with what appears to be one centrally located nucleolus (the latter is ultrastructurally complex in *Acrasis rosea*).

It has long been recognized that the amoebae of acrasids do not aggregate in the manner characteristic of many dictyostelids. They do not align themselves into streams leading into the aggregation centers but enter singly or in islets of cells. Also, they do not become elongated in the direction of movement as do aggregating dictyostelid amoebae, nor has any distinctive "stickiness," so characteristic of aggregating dictyostelid amoebae, been reported among the acrasids. Nothing is known about the nature of possible attractants leading to aggregation in this group, although it has been found that *Acrasis* amoebae do not respond to the dictyostelid attractant, cyclic AMP (T. M. Konijn, personal communication).

Recent investigations by light and electron microscopy have supported

E. W. Olive's conclusion that the dictyostelid fruiting body is more highly developed than that of the acrasids. One very obvious difference is that in dictyostelids the cells entering the stalk secrete walls and die, thus being sacrificed for a supportive function, whereas in acrasids the stalk cells remain viable. A second major distinction is found in the presence of a cellulosic stalk tube in the dictyostelid fruiting body (see Chapter 3) and its absence in acrasids (L. S. Olive, 1970). Developmental studies have demonstrated that this is not a matter of simple gain or deletion of a structure but is rather a reflection on basic differences in patterns of sporulation. In the Acrasea pseudoplasmodia do not become sluglike or migrate. There is no evidence of a sexual process in the life cycle.

II. OCCURRENCE, ISOLATION, AND MAINTENANCE

Like their dictyostelid counterparts acrasids are of widespread occurrence on a great variety of substrates. Some are found only on dung of various animals, including monkey, dog, ruminants, horse, pig, rodents, and birds. One has been reported only on tree bark and dead wood and another on moribund plant structures that have remained attached to the plant. A single interesting but poorly known species was found once on decaying beans and another only once on old beer yeast.

One of the most widely distributed species is *Acrasis rosea*, found on old attached pods, capsules, fleshy fruits, inflorescences, etc., throughout temperate and tropical zones around the world. It is best isolated by soaking bits of the substrate in distilled water for about half an hour, blotting them, and then placing them on a weakly nutrient agar (e.g., hay infusion agar, Appendix). The pink, arborescent fruiting bodies with branching spore chains are readily detected under a dissecting microscope, usually on the substrate but sometimes on the agar surface. *Acrasis rosea* is easily cultured on any of several yeasts on cornmeal–dextrose agar (see Appendix), under which conditions it grows and sporulates abundantly. Sporulating cultures stored at coldroom temperatures (e.g., 8°C) should be transferred monthly. Reinhardt (1966) has reported that cells of sorocarps stored in silica gel at 4°C survive as long as 13 months. He recently found (personal communication) that viable cells were still present after 6 years.

Pocheina rosea produces its minute, pink, capitate sorocarps on bark of coniferous and frondose trees. They appear on pieces of bark moistened with distilled water and placed in a moist chamber for 2–4 days. It is usually necessary to examine a considerable number of collections before the species can be found. On one occasion we observed excellent sporulation on a collection of white oak bark treated in this manner. Variations in the time required for sporulation in the laboratory appear related to weather conditions at the time the collection is made. For example, sporulation occurs more quickly in the laboratory if the bark is moistened by rain the day before its collection. All efforts to culture this interesting species in the laboratory have failed—another of the many examples of mycetozoans on bark that have proved difficult to culture. In this connection, it is interesting to note that our attempts to culture an isolate of *Acrasis rosea* found once on bark met with failure, in contrast to the usual ease with which this species may be grown.

Of the coprophilous species, *Guttulinopsis vulgaris* is the most common. It is especially frequent on cow and horse dung. The glistening whitish, short-stalked fruiting bodies generally appear within 3 or 4 days on fresh dung that has been moistened and placed in a covered glass bowl. Good growth and fruiting can be obtained by transferring a mucilaginous spore mass to a plate of dung–urea agar or Raper's tryptone–glucose–yeast extract agar (see Appendix), with *Aerobacter aerogenes* as the food organism. After a few transfers on these media, this cellular slime mold loses its ability to fruit normally, a condition that can be overcome by inoculating sterile cow dung with a suspension of cells from the culture. Normal sporulating cultures are best maintained at cool temperatures on sterile dung and transferred every 2 months.

Raper (1960) has discussed culture methods for several unusual acrasids isolated in his laboratory but not formally described. The culture of *Copromyxa arborescens*, a new species isolated from cow and horse dung, has recently been reported by Nesom and Olive (1972). In the presence of *Escherichia coli* it grows and fruits on sterilized cow dung and on malt–yeast extract agar (see Appendix). Much more information is needed on the distribution of acrasids, the nature of their substrates, and requirements of culture. No member of the class has been cultured axenically.

III. LIFE CYCLE OF AN ACRASID
(ACRASIS ROSEA)

The three species of acrasids that have been studied in some detail in the laboratory show three different modes of sorogenesis. What appears to be a fourth type has been found by Raper (1973) in an undescribed new genus and species. Therefore, it is not possible to say that

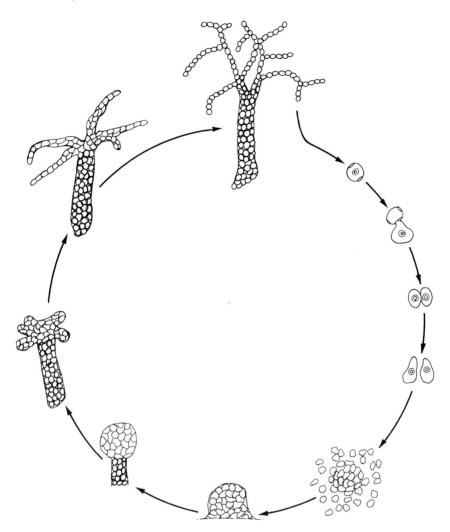

Fig. 165. *Acrasis rosea;* diagram of life cycle.

any single life cycle is representative of the group. *Acrasis rosea* (Fig. 165) is selected for discussion here because it is the species that we have studied in greatest detail (Olive and Stoianovitch, 1960; Olive *et al.*, 1961).

All cells of *A. rosea* have a pinkish to orange color caused by carotenoid pigments in the protoplasm. The spore, which germinates within about 1 hour in water or on agar, produces a single uninucleate amoeboid cell through an elongate opening in the wall. The amoebae (Figs.

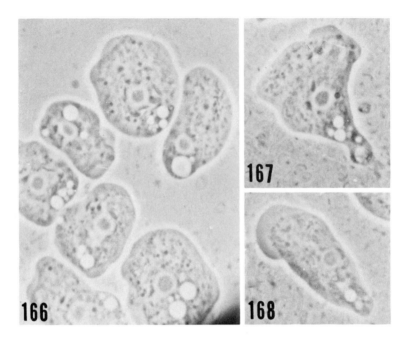

Figs. 166–168. *Acrasis rosea* amoebae. Fig. 166. Group of stationary amoebae. Figs. 167 and 168. Migrating amoebae. Note hyaloplasmic anterior margins and posterior uroid regions. Phase contrast ×1260.

166–168), described in Section I, move about by means of lobose pseudopodia, engulfing cells of the food organism or cannibalizing each other, now and then becoming quiescent and dividing by binary fission. Rapid multiplication occurs on cornmeal agar and a large population of cells, orange in mass, develops. Many of the amoebae become rounded, secrete walls around themselves, and develop into microcysts, often with numerous engulfed food cells in the protoplast.

Fruiting begins a few days after the culture is initiated. Although this occurs most abundantly when the food supply is depleted, sorocarps sometimes arise from areas with an abundant growth of the food organism. Most often the amoebae migrate in large numbers to the periphery of a colony, where they begin the aggregation phase of sorogenesis. The cells of acrasids are not elongated in the direction of movement and enter the aggregation centers not in streams but singly and in groups ranging from a few cells to broad sheets (Figs. 165, 193, and 195). The mature aggregates, which may be termed pseudoplasmodia (Fig. 169), are bright orange in color. They do not migrate across the substrate, although the larger ones may subdivide into two to several segments that then move apart for short distances. The pseudoplasmodium or each component of a subdivided pseudoplasmodium functions as a sorogen. The sorogen becomes surrounded by a thin slime sheath.

Under natural day and night conditions, sporulation in laboratory cultures (and probably in nature) occurs during the morning hours. During sorogenesis a capitate mass of amoebae rises at the apex of the developing stalk (Fig. 170). As new cells are added distally to the stalk by this sorogen, they become encysted. Their cell walls, in addition to the slime sheath surrounding the stalk, are primarily responsible for the supportive qualities of the stalk. Small fruiting bodies have

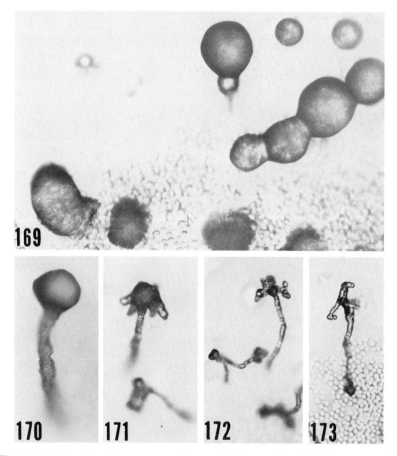

Figs. 169–173. *Acrasis rosea*, morphogenetic stages. Fig. 169. Sorogens, ×215. Fig. 170. Stalk and sorogen, ×110. Fig. 171. Lobed sorogen, ×110. Fig. 172. Sublobed sorogen, ×110. Fig. 173. Spore chain development, ×110. From Olive and Stoianovitch (1960).

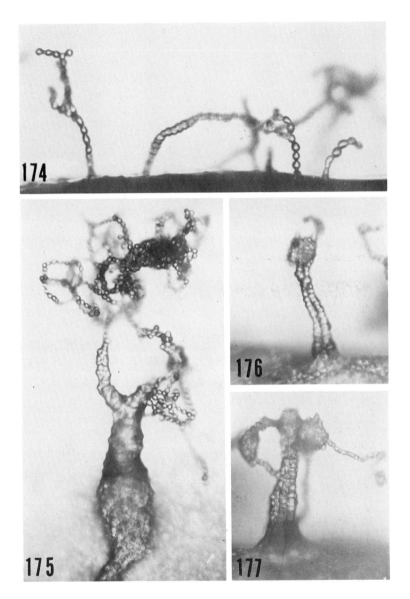

Figs. 174–177. *Acrasis rosea*, sorocarps. Fig. 174. Simple sorocarps with uniseriate stalks, ×280. Fig. 175. Large treelike sorocarp, ×130. Figs. 176 and 177. Fruiting bodies with parallel rows of stalk cells; Samoan isolate, ×200. Figs. 174 and 175 from Olive and Stoianovitch (1960).

stalks consisting each of a single row of cylindrical to rounded cells with their end walls closely appressed to each other (Fig. 174), whereas larger ones have trunklike stalks a few to many cells in thickness (Fig. 175). In the latter type, the stalk cells are parenchymatous in appearance, rounded, and with numerous intercellular spaces. In one *Acrasis* isolate (Figs. 176 and 177) the stalk cells frequently occur in parallel rows. Stalk cells of *Acrasis* do not become empty and dead as in dictyostelids but remain viable and germinable. In addition, no stalk tube has been found in *Acrasis* or any other member of this class. Sorogenesis is obviously a simpler process than in dictyostelid cellular slime molds.

When stalk formation is complete the sorogen at its apex enters into a most interesting series of developments leading to spore formation. If the sorogen is moderate to large in size, it first becomes lobed (Fig. 171). If small, it remains unlobed and develops into a single chain of spores, with or without branches, usually on a uniseriate stalk (Fig. 174). In the former case, if the lobes are small, each develops into an unbranched or branched spore chain. Larger sorogens commonly produce lobes that in turn become lobed again before producing similar spore chains (Fig. 172). The latter develop in a most unusual manner. The amoebae in a small sorogen or in a sorogen lobe move about in such a way as to arrange themselves into a single series (Fig. 173). As soon as the catenulate arrangement occurs the cells round up and encyst. Any group of amoeboid cells excluded along the side of a main developing chain may, in similar fashion, develop into a side chain. Thus, mature sorocarps show all gradations from simple uniseriate stalk and spore chain to large arborescent structures with trunklike stalk and numerous branching chains of spores (Figs. 174–177). In our investigations of *A. rosea* we have observed no development comparable to that briefly noted by van Tieghem (1880) in *A. granulata;* i.e., that amoebae migrate from base to top of the mature stalk and there become superimposed to form chains of spores.

Although both stalk cells and spores may serve as propagules, producing one amoeba each at germination, they are morphologically distinct. The globose to subglobose spores are distinguished by the presence of hila, which appear as small rings at the points of spore contact (Fig. 178). Therefore, each spore in a chain, except the terminal one, has two hila unless a secondary chain is attached to it, in which case it has three. The cell at the base of a chain is transitional, having a hilum distal to but none proximal to the stalk. It is obvious that spore differentiation is influenced not only by the uniseriation of cells but also by position within the developing fruiting body, since a simple sorocarp that is uniseriate throughout has spores with and stalk cells without

Fig. 178. *Acrasis rosea;* drawing of spore chains showing hila. Spores on left have germinated, ×980. From Olive and Stoianovitch (1960).

hila. Spores are further distinguished from stalk cells in being dispersable by air currents. Their separation from the fruiting body is facilitated by their hilar connections. The sudden opening and closing of a culture dish creates air movements sufficient to cause detachment of the spore chains, which thereafter may be found scattered over the agar surface. Since *Acrasis rosea* occurs on dead attached plant parts and the sorocarps mature during the morning hours when dew is likely to be present on these structures, the spores are more likely to be distributed by wind at a time suitable for their germination.

The original report that the type collection of *A. rosea* requires light for fruiting (Olive and Stoianovitch, 1960) has been confirmed by Reinhardt (1968a) for this and most other isolates of this species. Reinhardt has also discovered that a minimum dark period is required by these isolates following illumination. In NC-18, an isolate that sporulates profusely, he found no fruiting in cultures maintained in either uninterrupted light or dark incubation. Dark-grown cultures subjected to short or lengthy periods of illumination fruited when returned to the dark. The first signs of aggregation appeared about 6 hours later; numerous well-developed aggregation centers were present at 8 hours; and within 12–14 hours all stages in sorogenesis, including mature sorocarps, were present. The minimum dark period for the production of even a few sorocarps was 7–8 hours. If cultures were illuminated by blue light (400–475 nm) for as little as 10 minutes after 6 hours of dark incubation, the aggregation centers broke up and fruiting was delayed for 4 or 5 hours. If left in continuous light after the 6-hour dark period they failed to fruit altogether. Red and far red light are ineffective inhibitors of fruiting. Reinhardt suggested that interruption of the essential dark period by light may destroy an aggregation inducer or inhibit its production. Nevertheless, the initial light exposure is probably a prerequisite for synthesis of the inducer. Neither amoebae nor sorocarps become oriented toward a unidirectional light source.

Since the time of fruiting can be predetermined by regulating the start of dark incubation following exposure to light, it is likely that the natural cycle of development in *Acrasis rosea* is controlled by normal day and night periods rather than by an inherent circadian rhythm. The nature of a possible photoreceptor system influencing sorogenesis is unknown. Fuller and Rakatansky (1966) reported that three different carotenoids are present—the xanthophyll torulene, an unidentified xanthophyll, and an unidentified carotene. All proved to have absorption maxima in the blue wavelengths, which are also the most effective in stimulating fruiting (Reinhardt, 1968a).

IV. CLASSIFICATION

Apparently, the first acrasid to be described was *Guttulina rosea* Cienkowsky (1873), until recently known only from a single collection on dead wood in Russia and described in an obscure report. Following this, van Tieghem (1880) described *Acrasis* and placed it along with *Guttulina* (and *Dictyostelium*) in his new order Acrasieae. Another genus was added when Zopf (1885), for reasons to be given below, changed the name of *Guttulina protea* Fayod (1883) to *Copromyxa protea* (Fayod) Zopf. The latter generic name was later accepted by Nesom and Olive (1972). In his monograph of the cellular slime molds, E. W. Olive (1902) described *Guttulinopsis,* the last genus of acrasids to be described. He accepted van Tieghem's order Acrasieae and created the new family Guttulinaceae to include *Guttulinopsis* and *Guttulina,* retaining Fayod's *G. protea* in the latter genus. Since *Acrasis granulata* van Tieghem was (and still is) known only from its original description and its author had failed to make clear whether limax- or eumycetozoan-type amoebae were present, its relationships to other cellular slime molds were obscure, and E. W. Olive chose to classify it among the dictyostelids.

After the discovery of *Acrasis rosea* Olive and Stoian. (1960) and the demonstration that it has limax amoebae, the genus was transferred to the Guttulinaceae. A year later, however, Loeblich and Tappan (1961) reported that the generic name *Guttulina* had been first applied to a protist unrelated to *G. rosea,* the type species of the acrasid genus. They therefore proposed the generic name *Pocheina* (in honor of the protozoologist Poche) and the new combination *P. rosea* (Cienk.) Loebl.

and Tapp. for the type species. However, the family name Pocheinidae of Loeblich and Tappan proved superfluous in view of the fact that the name Acrasidae Poche (1913) remains a valid taxon for accommodating *Pocheina* and *Acrasis*.

In a recent revised classification of the Mycetozoa (L. S. Olive, 1970), the subclass Acrasia, coordinate with the subclasses Protostelia, Dictyostelia, and Myxogastria, was erected. Two acrasid families were recognized, the Acrasidae with *Pocheina* and *Acrasis* and the Guttulinopsidae (a new family) with the single genus *Guttulinopsis*. E. W. Olive had originally distinguished the latter genus from *Guttulina* on the basis of its having pseudospores rather than true spores. The pseudospores were characterized as "encysted individuals without cell walls," which during germination reverted to the amoeboid state without leaving empty spore cases. However, it was later discovered that thin spore walls are present in *G. vulgaris* E. W. Olive, the type species, although they often appear to be resorbed during germination (L. S. Olive, 1965). The genus is nonetheless distinguishable in several ways from other acrasids.

A number of acrasid cellular slime molds that have not yet been described, certain of which probably require the recognition of new genera, are being studied by Raper (1960, 1973). The sorocarps of some are hardly more than unbranched columns of encysted cells which, as in *Copromyxa*, are all of one type throughout the sorocarp. *Copromyxa* itself has certain structural and developmental features that make it out of place in either of the two presently recognized families. The unique features of the genus have led us to create the new family name Copromyxidae to accommodate it and future taxa with similar characteristics.

Key to the Families and Genera of Acrasida

Cells of sorocarp differentiated into morphologically distinct spores and stalk cells; flagellate cells present or absent .... **Family Acrasidae**
    Spores in a terminal spherical or subspherical sorus; biflagellate cells produced....................... *Pocheina*
    Spores produced in simple or branched chains; flagellate cells absent............................. *Acrasis*
Cells throughout sorocarp all essentially alike; flagellate cells absent
    Sorocarps arising directly from substrate; varying from unbranched and columnar to arborescent.......... **Family Copromyxidae**
                                Single genus: *Copromyxa*
    Sorocarps capitate, arising from pseudoplasmodial plaques; spores contained in membranous compartments .... **Family Guttulinopsidae**
                                Single genus: *Guttulinopsis*

V. DESCRIPTIONS OF TAXA

A. Acrasidae Poche (1913)

The trophic stage is primarily comprised of uninucleate amoeboid cells, although anteriorly biflagellate cells are known in one genus. The pseudoplasmodia produce sorocarps singly or become subdivided and form several. The sorocarp is stalked and bears either a capitulum of spores or branched or unbranched chains of spores at its summit. The stalk cells are viable and germinate readily. Spores are distinguishable from stalk cells by the presence of hila (contact scars) on their walls. Two genera are known.

1. *Pocheina* Loeblich and Tappan (1961)

The type and only known species of the genus is *P. rosea* Loebl. and Tapp. Because *Guttulina* Cienkowsky (1873) was antedated by *Guttulina* d'Orbigny, a foraminiferan genus, recognition of a new generic name for the acrasid has been necessary. *Pocheina rosea* appears to have been found only once by Cienkowsky, who collected it on dead lichenized wood in Russia. Although E. W. Olive (1902) included it in his monograph without having seen it, there is no evidence of its having been collected again until recently. K. B. Raper (1973) reported its occurrence on coniferous bark from the Netherlands (Fig. 181), but he was unable to culture it or to germinate the spores.

During the past 2 years we have repeatedly found what we conclude to be *P. rosea* on bark of various trees (larch, pine, wild cherry, oak, maple) placed in a moist chamber for 1–4 days (Figs. 179 and 180). The spores and stalk cells of several collections have been induced to germinate on lactose–yeast extract agar and oak bark agar (see Appendix) at pH 5.1, with germination beginning in some cases within 30–60 minutes. At this time flagellate cells were observed and two types of germination noted. When spores or stalk cells germinate on agar in the absence of excess water, each gives rise to a single amoeba with lobose pseudopodia (Fig. 182) that crawls about very actively, often revealing a posterior uroid region with filose extensions, as in *Acrasis rosea*. However, if sufficient water is present the spores (and stalk cells) may germinate in two ways—in the manner just described and also to produce flagellate cells. In the latter event, a nuclear division occurs just before emergence of the protoplast, and the resultant binucleate cell develops two pairs of flagella before or soon after it leaves the spore case (Fig. 183). The cell then divides into two uninucleate planonts (Fig. 184), each with an anterior pair of equal flagella (Figs. 185 and

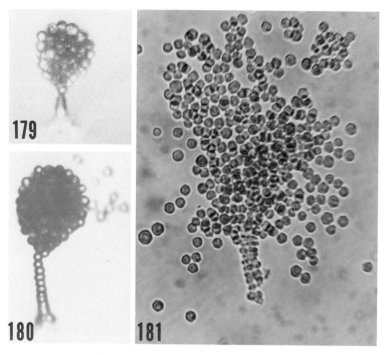

Figs. 179–181. *Pocheina rosea,* sorocarps. Figs. 179 and 180. Sorocarps with single tier of stalk cells, ×325. Fig. 181. Sorocarp mounted in water, showing more than one tier of stalk cells and dispersed spores, ×340. Figs. 179 and 180 from Olive and Stoianovitch (1974); Fig. 181 from K. B. Raper (1973).

186). It seems likely that the nuclear division in the spore activates the centriole–kinetosome apparatus that leads to flagellation. The planont is somewhat short to elongate, with a pointed posterior end. There is some evidence of a spiral twist in the body and longitudinal surface ridges. The cell swims about rapidly, rotating on its axis. Unfortunately, we have not had sufficient numbers of these cells on hand at one time to allow a detailed investigation of the flagellar characteristics, nor has it been possible to culture this bark-inhabiting species in the laboratory, in spite of exhaustive attempts.

A description of *Pocheina rosea,* based largely on our collections, is as follows: sorocarps pinkish orange, surrounded by a thin slime sheath, 98–150 μ tall; sorus globose to obovate, 47–105 μ in diameter; stalks 31–100 μ long, comprised of smooth, nonareolate cells in one to several rows; spores globose, finely punctate, areolate, 8.7–12.5 μ in diameter; spores and stalk cells germinating to produce each a single uninucleate amoeboid cell with lobose pseudopodia or two anteriorly biflagellate cells; occurring on dead lichenized wood (type collection) and bark of coniferous and frondose trees in Russia, the Netherlands, and the United States (North Carolina, Wisconsin, and Michigan). There is no evidence that a type collection still exists.

Pocheina rosea is of major interest because it is the first cellular slime mold in which flagellate cells have been found. The consistent presence of two equal flagella and the apparently more highly modified sorocarp further emphasize the differences between acrasid and dictyostelid cellular slime molds.

Two other species, known only from their original collections in France were placed in the genus *Guttulina. Guttulina aurea* van Tieghem (1880) was described as having stalked fruiting bodies similar to those of *P. rosea* but golden yellow in color and with spherical spores 4–6 μ in diameter. It was collected on horse dung. The other, *G. sessilis* van Tieghem (1880), was described as producing white stalkless fruiting

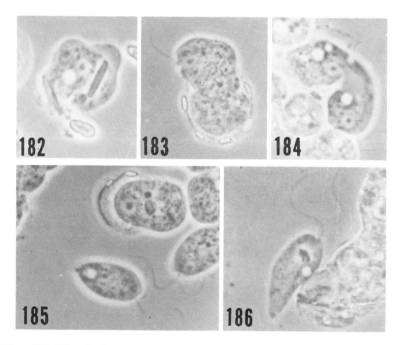

Figs. 182–186. *Pocheina rosea*, spore germination. ×1260. Fig. 182. Uninucleate amoeboid cell from spore. Note lobose pseudopodia and ingested conidia. Fig. 183. Flagellate (binucleate) protoplast arising from spore (note spore areoles). Fig. 184. Binucleate flagellate protoplast from spore dividing by plasmotomy. Fig. 185. Germinating spore and uninucleate biflagellate cells. Fig. 186. Uninucleate biflagellate cell. Figs. 182, 184, and 186 from Olive and Stoianovitch (1974).

bodies on decaying beans. The oval spores measured around 4×8 μ. In both species the spores were contained in a mucilaginous matrix. E. W. Olive (1902) considered the possibility that these organisms might belong in the genus *Guttulinopsis*. Since they have not been collected again and so little is known about them, their real affinities remain uncertain. In any event, there is little to indicate a close relationship with *Pocheina* or *Acrasis*. It is highly desirable that the species be reisolated and studied more carefully in the laboratory.

2. *Acrasis* van Tieghem (1880), emend. L. Olive (1970)

a. Taxonomy. The type species of the genus, *A. granulata* van Tieghem, is known only from the original collection on beer yeast. It proved to be markedly different from previously described cellular slime molds in having the spores borne in simple chains rather than in globose heads. The stalk cells were described as being arranged in a single row with a cramponlike cell at the base (a characteristic that may have been emphasized by a basal flare in the surrounding sheath). Sorogenesis was shown to be preceded by aggregation of amoebae. In more vigorous cultures up to 10 or 12 simple uniseriate fruiting bodies frequently became united into coremium-like groups. The spores were described as having minutely punctate walls with a brownish-violet pigmentation that caused the fruiting bodies to appear blackish to the unaided eye. Unfortunately, van Tieghem failed to make clear whether the amoebae had filose or lobose pseudopodia, and the presence of limax-type amoebae can only be inferred from comparisons with *A. rosea,* the only other cellular slime mold known to have catenulate spores.

Acrasis rosea is a relatively common and widespread cellular slime mold that has been found on a great variety of dead plant parts that have remained attached to the plant. The simplest fruiting bodies resemble those of *A. granulata* in being uniseriate, but most sorocarps have branching chains of spores and many have trunklike stalks (Figs. 174 and 175). Except for the hila or areolae at their points of contact, the globose to subglobose spores, which measure $7.2–13.5 \times 8.1–15.3$ μ, are smooth-walled and without wall pigmentation. The pink to orange color of all cells of this species is traceable to carotenoids in the protoplasm. The sorocarps have a bright pink color that is obvious to the unaided eye. In one isolate the larger stalks were found to be comprised of rows of cells (Figs. 176 and 177), a condition which approaches that of the coremiform fruiting bodies of *A. granulata.*

The species has a wide range of utilizable food organisms, including ascomycetous and basidiomycetous yeasts, conidia of *Phoma* and *Colletotrichum, Flavobacterium* sp., and species of the algal genera *Chlorococ-*

cus and *Stichococcus* (Olive and Stoianovitch, 1960; Olive *et al.*, 1961; Reinhardt, 1968a).

The type collection of *A. rosea* fails to grow on the yeast *Hansenula anomala* alone. Weitzman (1962) found that growth and fruiting occur in the presence of this yeast on Sabouraud's agar if a cell-free filtrate from washed cells of *Rhodotorula mucilaginosa* is added. Growth also occurs on a mixture of *Hansenula* and *Rhizobium* sp. or any of several other unidentified gram-negative, rod-shaped bacteria, none of which alone supports growth. Filtrates of two of these bacteria in combination with *Hansenula* also permit growth of the amoebae. There were no permanent adaptations to the yeast. These experiments demonstrate the usefulness of water-soluble diffusible substances in the nutrition of a cellular slime mold and suggest a complexity in nature that is not commonly detected in the laboratory.

b. **Variation in *Acrasis rosea*.** Five isolates from different localities showed individual differences in food preferences when tested on eight different microorganisms in culture (Olive *et al.*, 1961). The type collection and NC-18, which were found to differ sharply in their response to three of these food organisms, were selected for further study. In addition, NC-18 was a more vigorous isolate with larger fruiting bodies. Both isolates had certain food organisms in common, including *Rhodotorula mucilaginosa* and *Phoma conidiogena*, on which they grew and fruited well. However, the type, but not NC-18, was able to grow and sporulate on *Flavobacterium* sp., whereas NC-18, but not the type, grew and sporulated on *Hansenula anomala* and on *Colletotrichum lagenarium*. When these two races were mixed and cultured together on the same plate and single-amoeba or single-spore isolates were later made, these yielded neotypes, able to grow on a different combination of test food organisms, at the rate of about 4% and 3%, respectively. Mixed fruiting bodies from which both races could readily be isolated were found in the original mixed plate. In successive transfers the neotypes reverted to type or to NC-18 but not to a mixture of the two, indicating that they did not represent genetic recombinants. The observation that cells of *Acrasis rosea* anastomose freely under certain conditions provides a possible explanation for the origin of these neotypes. They are possibly heteroplasmons in which the altered characters have resulted from an exchange of cytoplasmic organelles and macromolecules, which are eliminated in subsequent cell divisions and subculture.

Dutta and Garber (1961) found that races of the plant pathogen *Colletotrichum lagenarium* could be identified by the degree of susceptibility of their conidia to ingestion by six different races of *Acrasis rosea* on four different culture media. The procedure is somewhat comparable to the typing of pathogenic bacterial strains with bacteriophages. In a later experiment (Garber and Dutta, 1962), it was found that *A. rosea* amoebae could ingest conidia of race 2 of the pathogen but not of race 1. The results further indicated that uninucleate conidia from heterokaryons between the two fungal races, apparently containing only race 1 nuclei but mixed cytoplasm, were ingested by the amoebae.

Conspicuous differences among isolates of *A. rosea* have also been observed with respect to their fruiting responses to light (Reinhardt, 1968a). For example, only five out of 15 isolates from various parts of the world were able to produce some sorocarps when grown in continuous light. Variations were also found to exist with regard to the minimal dark periods for optimal fruiting. Although there is no evidence of sexual reproduction in *Acrasis* or any other acrasid, the possibility of a parasexual process has not been eliminated. If parasexuality should occur in *A. rosea*, then the numerous phenotypic differences among isolates could prove useful in genetic experiments.

B. Copromyxidae Olive and Stoianovitch, *Fam. Nov.**

The family is distinguished by some of the simplest sorocarps found in the cellular slime molds. All cells in a sorocarp are morphologically similar and viable. Because there is no differentiation into spore and stalk cells, all encysted cells of the fruiting body are referred to here as "sorocysts." The limax-type amoebae resemble those of *Acrasis rosea* but are unpigmented. Flagellate cells are absent. The single known genus, *Copromyxa*, contains only two described species. It is probable that certain undescribed isolates of acrasid cellular slime molds under study in Dr. K. B. Raper's laboratory (Raper, 1960, 1973) will eventually find a place in this family.

Copromyxa Zopf (1885)

Fayod (1883) found on dung of horse and cow in Germany a cellular slime mold with limax-type amoebae which he named *Guttulina protea*. It was described as having simple to much-branched sorocarps 1–3 mm high comprised throughout of colorless to slightly yellowish, oval to bean-shaped or triangular "spores" (sorocysts) that measured 9×14

---

* Copromyxidae L. Olive and Stoianovitch, *fam. nov.*

Amoebae pseudopodiis lobatis praeditae; pseudoplasmodia nonmigrantia; sorocarpia columnaria vel arborea; loculi membranacei absens; sorocystae in sorocarpiis ubique similaria et sine areolis.

μ. Spherical microcysts, 12–15 μ in diameter, with roughened gold to brown walls were also described. Apparently, no organism with this combination of characteristics has been found since the original collection. Zopf (1885), noting that the type species of *Guttulina* showed a morphological distinction between spores and stalk cells, decided that *G. protea* had been wrongly classified in that genus. He therefore erected the new genus *Copromyxa* to accommodate it, calling it *C. protea* (Fayod) Zopf. However, E. W. Olive (1902) preferred to retain the species in the genus *Guttulina*.

In recent years we have repeatedly found on cow dung collected in North Carolina a cellular slime mold that resembles Fayod's species but differs in at least one important character—the microcysts are indistinguishable from the sorocysts under both light and electron microscopes (Nesom and Olive, 1972; Nesom, 1973). A somewhat similar isolate from horse dung in Wisconsin, sent to us by Dr. K. B. Raper, was concluded to be the same species, *Copromyxa arborescens* Nesom and Olive (1972). The sorocarps are whitish, simple and horn-shaped to branched and treelike (Figs. 187 and 188). The sorocysts (Fig. 189) are smooth walled, hyaline, and oval, oblong, bean-shaped, or triangular in outline, measuring 4.4–12.5 × 7.5–17.5 μ. They never have the uniseriate arrangement characteristic of *Acrasis* spores and hila are absent from their walls. Within the sorocarp there is no distinction between encysted cells in the stalklike region and in branches arising from its summit. The microcysts (Fig. 190), which develop directly on the substrate, resemble the sorocysts in every feature, including shape and size. There is a distinct possibility that the microcysts of Fayod's species were actu-

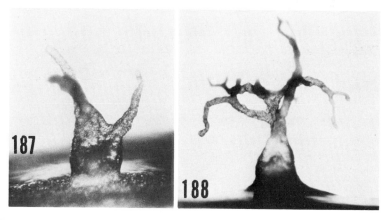

Figs. 187 and 188. *Copromyxa arborescens*, fruiting bodies, ×56 and ×45, respectively. From Nesom (1973).

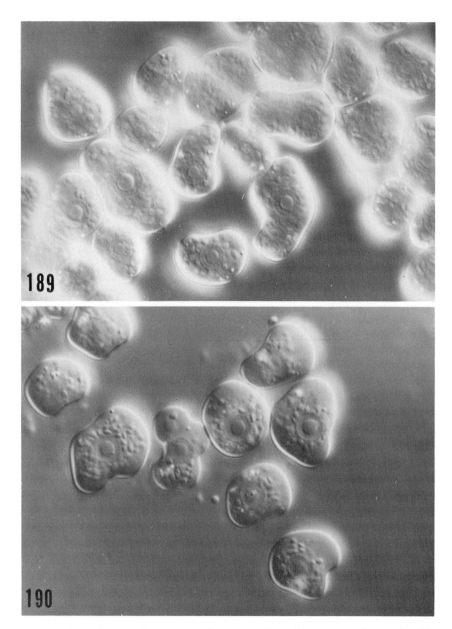

Figs. 189 and 190. *Copromyxa arborescens,* Nomarski contrast micrographs. Fig. 189. Sorocysts, ×1250. Fig. 190. Microcysts, ×1250. Courtesy of M. Nesom.

ally encysted cells of another amoeboid organism. If so, the species name, based on the description of two different species, is not valid.

Copromyxa arborescens grows and fruits well on sterilized cow dung or on weak malt–yeast extract agar in the presence of bacteria as the live food source. The amoebae move about by means of lobose pseudopodia in a manner similar to that in *Acrasis rosea* (Fig. 191). Each amoeba typically has a single centrally uninucleolate nucleus and one

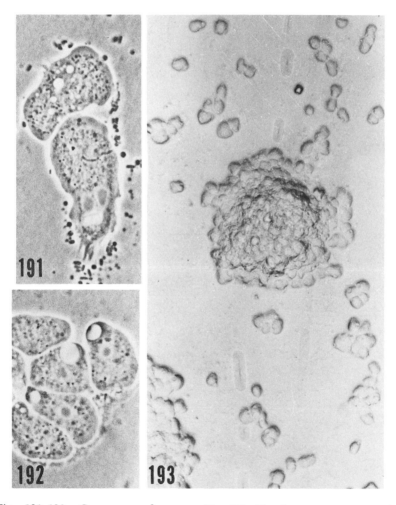

Figs. 191–193. *Copromyxa arborescens*. Fig. 191. Binucleate migrating amoeba, showing posterior uroid region with filose extensions, ×1800. Fig. 192. Cluster of stationary uninucleate amoebae. Note contractile vacuoles. ×1400. Fig. 193. Aggregation of amoebae, ×170. From Nesom (1973).

contractile vacuole (Fig. 192). In an actively migrating amoeba the contractile vacuole is located in the vicinity of a posterior uroid region with sticky filose extensions. Up to eight nuclei have been found in some cells. Plasmotomy occurs among the plurinucleate cells, whereas mitotic cell division is of the usual binary type.

In the manner characteristic of acrasid cellular slime molds, aggregation does not result from cell streaming but from the migration of amoebae singly and in irregular groups into the aggregation centers (Fig. 193). The aggregates eventually become doughnut-shaped by a peripheral shifting of the amoebae. These pseudoplasmodia produce from one to several sorocarps. Fayod described a method of fruiting, now confirmed for *C. arborescens* (Nesom, 1973), that appears to be unique among the cellular slime molds, with the possible exception of van Tieghem's report on the poorly known *Acrasis granulata*. As amoebae encyst to form the basal part of the sorocarp, other amoebae crawl to its top and encyst, thus increasing its length. Apparently the branches of the sorocarp develop in the same way, for if cells of *C. arborescens* with the capacity to aggregate are stained with neutral red and placed around the aggregate at the base of a developing sorocarp, reddish groups of cells later appear at the developing branch tips (Nesom, 1973). It will be of interest to learn whether a cellular slime mold tentatively labeled *Guttulinopsis* by K. B. Raper (1960, pp. 387–388 and Figs. 24–30) has a similar mode of sorogenesis. Its sorocarps consist of upright columns of encysted cells that also show no differentiation into stalk cells and spores.

The apparent absence of a true sorogen in *Copromyxa* and the lack of differentiation among the encysted cells of the sorocarp indicate a less advanced type of sorogenesis than that reported in other cellular slime molds. Obviously, definitions of "sorogenesis" and "sorocarp" must be extended here if they are to include such forms as *Copromyxa* because there is some question as to whether a true sorogen or sorus is present in this genus. These characteristics also raise the question of whether the genus is related to *Pocheina* and *Acrasis* or whether cellular slime molds with limax-type amoebae have developed by more than one evolutionary line from simpler protists. Detailed studies of other acrasid cellular slime molds are required before such questions can be intelligently discussed.

C. Guttulinopsidae L. Olive (1970)

Raper (1973) follows E. W. Olive in placing both *Guttulinopsis* and *Guttulina* (*Pocheina*) in the family Guttulinaceae Zopf. However, detailed studies of *Guttulinopsis vulgaris* E. W. Olive, now considered the type species of the genus, reveal a type of sorocarp so different from that of *Pocheina* (*Guttulina*) *rosea* that the two can hardly be considered more than remotely related (L. S. Olive, 1965). Comparisons are further complicated by the discovery of a flagellate stage in *Pocheina* that at the same time is characterized by a more advanced type of sorocarp. Therefore, in the present treatment, *Pocheina* has been placed in the Acrasidae along with *Acrasis*, to which it appears closely related, leaving *Guttulinopsis* E. W. Olive (1901) as the only known genus of the newly erected family Guttulinopsidae (L. S. Olive, 1970). The latter is characterized by having amoebae with lobose pseudopodia and a nonstreaming aggregation pattern like that of *Acrasis* and *Copromyxa*. Flattened pseudoplasmodial aggregates that do not migrate are formed and these give rise to the sorocarps. The spores develop by encystment of amoebae within membranous compartments of the sorocarp.

Guttulinopsis vulgaris E. W. Olive (1901) is the only species that has been studied in any detail (L. S. Olive, 1965). It is a common and widely distributed species that occurs on dung of various types but most often on that of cow and horse. It has been found throughout much of the United States, including Hawaii, and in the West Indies, Mexico, Ecuador, Tahiti, American Samoa, Fiji, and Singapore. The species is readily collected by placing fresh dung of cow or horse in a moist chamber and wetting the surface. If the organism is present the sorocarps generally appear within 2 or 3 days as glistening white droplets on the substrate surface. Examination under the dissecting microscope reveals that each droplet is actually a mucous sorus of spores on a short, thickened stalk, with or without one or more small secondary sori (Fig. 194).

When the tip of a needle is applied to a sorus, most of it adheres and can be transferred to a culture plate. Media such as tryptone–glucose–yeast extract agar or fresh dung–urea agar in conjunction with live cells of *Aerobacter aerogenes* are most suitable for growth. Spore germination is usually not abundant but may be quite good on fresh dung–urea agar in which the odor of ammonia can be detected. A thin spore wall has been demonstrated but this appears to be resorbed during germination. The presence of a spore wall nullifies E. W. Olive's characterization of the genus as having "pseudospores" (resting cells without walls). The spore is uninucleate and produces a single uninucleate amoeba with contractile vacuole. There is a single central nucleolus. Multiplication of amoebae is rapid, and fruiting may occur within 2 or 3 days in culture. However, after several transfers the sorocarps become more irregular and the spores fewer and less normal in appearance. The culture may generally be restored to normal by transferring it to sterilized dung.

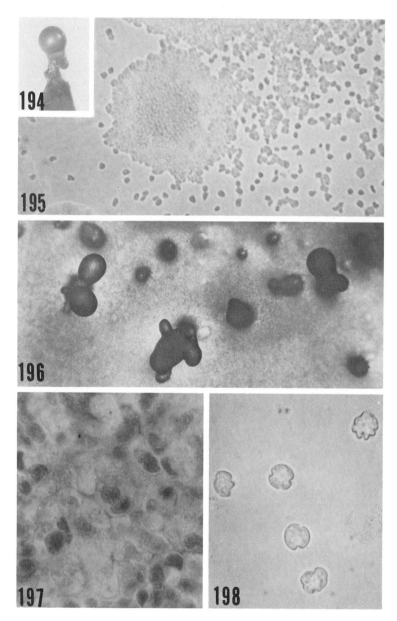

Figs. 194–198. *Guttulinopsis vulgaris.* Fig. 194. Sorocarp on dung. Note small lateral sorus, ×45. Fig. 195. Aggregation, ×125. Fig. 196. Sorocarps on agar surface, ×45. Fig. 197. Section of stalk showing cells in membranous compartments, ×900. Fig. 198. Spores, ×900. From Olive (1965).

During aggregation the amoebae move into the centers singly and in small groups. The amoebae are not elongated in the direction of migration (Fig. 195). The mature aggregate or pseudoplasmodium is a discrete mass of cells surrounded by a mucilaginous matrix and may be readily removed from the substrate as a unit. Most often one pseudoplasmodium gives rise to a single sorocarp. The details of sorogenesis are not known but a fruiting body with a relatively thick, short stalk and single terminal sorus, with or without subtending secondary sori, is formed (Fig. 196). Sections through maturing sorocarps show them to be partitioned into inconspicuous membranous compartments in which the amoebae encyst and become spores (Figs. 197 and 199). The upper

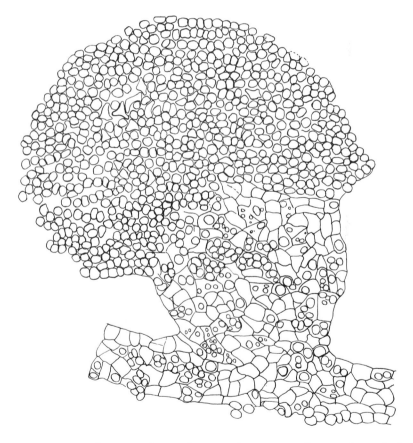

Fig. 199. *Guttulinopsis vulgaris;* drawing of sorocarp and part of pseudoplasmodial plaque in section. Note membranous compartments in stalk and pseudoplasmodium, ×340. From Olive (1965).

compartments of the sorocarp contain numerous spores that are released in a mucous mass, whereas the compartments of the stalk and pseudoplasmodium are more persistent and contain relatively few encysted cells. Many of the amoebae in the latter two areas appear to abort. Sporelike encysted cells may also develop in pseudoplasmodia that fail to produce sorocarps. The term "sorocyst" might be found more appropriate for all of these encysted cells if ultrastructural studies prove them to be alike. Microcysts do not seem to be present in this species.

Guttulinopsis vulgaris is most readily identified by its whitish sorocarps that are mostly 100–450 μ tall, with terminal sori 75–350 μ broad, and by the unique appearance of its spores. The latter are smooth walled and measure 2.6–7.8 × 3.9–9.2 μ. They are typically discoid or irregularly spherical with one to several flattened or concave sides and often exhibit several areas in which the protoplast seems to have drawn away from the spore wall (Fig. 198), very much as E. W. Olive described them.

Two other species of *Guttulinopsis,* described as *G. stipitata* E. W. Olive (1901) and *G. clavata* E. W. Olive (1901), both on dog dung, are known only from their original descriptions. There is insufficient information available to determine whether they are valid members of this genus. The undescribed "*Guttulina* sp." with round spores discussed by Raper (1960) appears very likely to be a species of *Guttulinopsis.* The "spongelike matrix" found within the sorocarp suggests a membranous compartmentalization like that of *G. vulgaris.* More detailed morphogenetic and taxonomic studies of this group are needed.

VI. ULTRASTRUCTURE

A. *Acrasis rosea*

The fine structure of *A. rosea* was studied by Hohl and Hamamoto (1968, 1969b). The protoplasts contain most of the common eukaryotic cell organelles (Figs. 200 and 203). Both contractile and food vacuoles were observed, along with distinctive structures called "P bodies" that were thought to be the pigment granules of this orange-colored organism (Fig. 201). Food vacuoles containing yeast cells or cannibalized *Acrasis* cells are common in amoebae and microcysts but not in cells of the mature fruiting body. The mitochondria are usually surrounded by cisternae of rough ER and contain cristae in the form of flat circular discs (Figs. 203 and 204), unlike the tubular cristae of *Copromyxa* and the dictyostelids. A recent discovery in our laboratory (Michael Dykstra, unpublished data) has revealed that mitochondria in the spores of

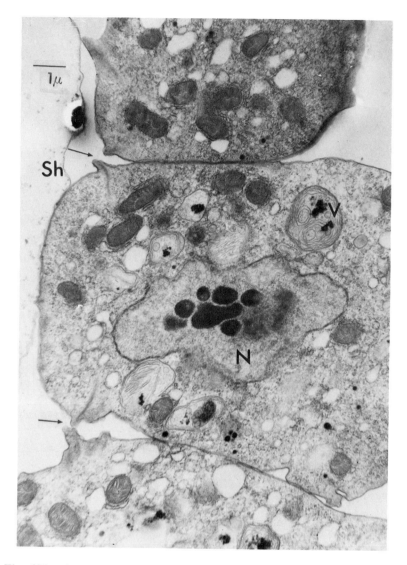

Fig. 200. *Acrasis rosea,* electron micrograph of series of cells in developing sorocarp. Note nucleus (N) with several nucleolar bodies, vacuole (V) with lamellar contents, slime sheath (Sh), and pseudopodium-like cell extensions with microfibrillar bundles (arrows). The latter may function in maintaining cell balance. From Hohl and Hamamoto (1969b).

Pocheina rosea, a species possibly related to *Acrasis rosea,* also have discoid cristae.

The nucleolus of *A. rosea,* which under the light microscope generally

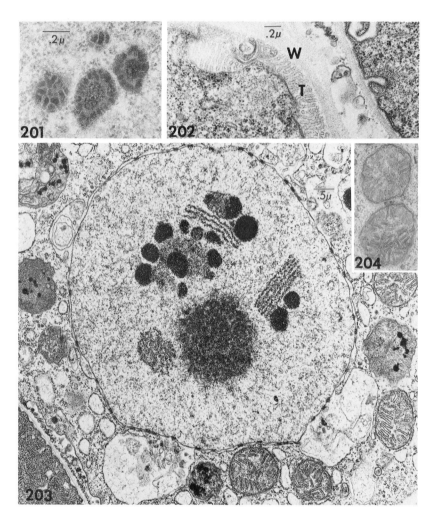

Figs. 201–204. *Acrasis rosea,* ultrastructure. Fig. 201. "P bodies," thought to be pigment granules, found in cells at all stages. Fig. 202. Periphery of differentiated cell in sorocarp, showing single wall layer of fibrillar material (W) and underlying layer of tubular structures (T). Fig. 203. Details of nucleus, showing lamellar structures and darker and lighter components of the complex nucleolus. Note also mitochondria with disc-shaped cristae. Fig. 204. Detail of mitochondria, showing disc-shaped cristae. Figs. 201 and 202 from Hohl and Hamamoto (1969); Fig. 203 courtesy of H. Hohl; Fig. 204 courtesy of D. J. Reinhardt.

looks like a single but often irregular body, is ultrastructurally more complex (Hohl and Hamamoto, 1968). It contains three morphologically distinct components: (1) a number of separate or aggregated round bodies with densely compacted ribosome-like particles, (2) occasional dense homogeneous bodies within the foregoing components and (3) a relatively large mass of homogeneous and less dense granular material. In encysted cells of the fruiting body an additional unusual structure, the "lamellar element," was found (Fig. 203). These elements, which occur singly or in stacks of up to 12, are layers of ribosome-like granules that are sometimes found to be continuous with the first type of nucleolar component. Histochemical tests indicate that the granules in the lamellae are either ribosomes or ribosome precursors.

In the pseudoplasmodium, and less commonly in the sorogen, conspicu-

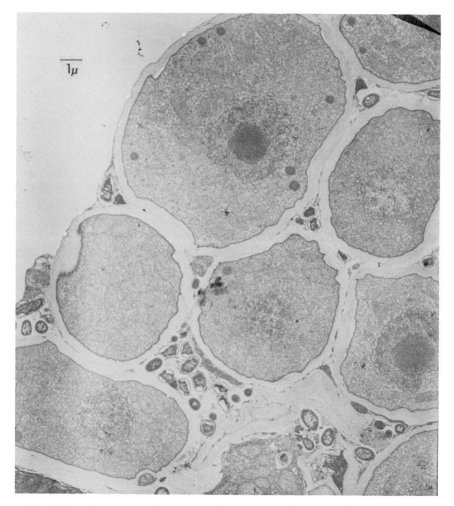

Fig. 205. *Copromyxa arborescens;* ultrastructure of sorocarp, showing sorocysts with unilayered walls of fibrillar material and nuclei with irregular envelopes. Note bacterial cells incorporated among the cysts. From Nesom (1973).

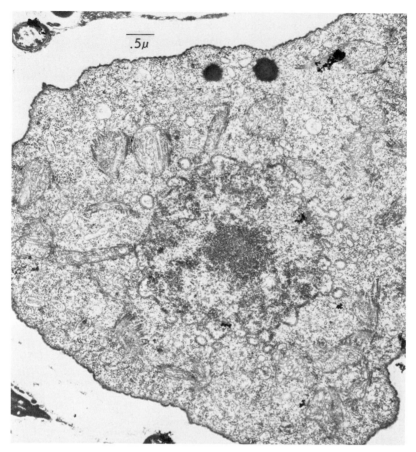

Fig. 206. *Copromyxa arborescens;* ultrastructure of sorocyst, showing numerous evaginations of nuclear envelope. Note mitochondria with tubular cristae. Surrounding wall is indistinct. From Nesom (1973).

ous bundles of microfibrils traverse large parts of the cell and often extend into pseudopodia (Fig. 200). They make contact with the plasmalemma and are sometimes found opposite each other in adjacent cells. They probably aid in the movement of amoebae during sorogenesis.

All encysted cells of *A. rosea* appear to have a simple wall consisting of a single layer of fibrillar material (Fig. 202). Lomasome-like contorted tubular structures are commonly present in peripheral pockets or broadly distributed between cell wall and plasmalemma. Hohl and Hamamoto (1969b) state that no differences were observed between stalk cells and spores. Most of their illustrations of encysted cells in the sorocarp show relatively broad contacts among cells that do not appear to be in uniseriate arrangement and no evidence of the hila that characterize spore connections. This indicates that they were dealing only with stalk cells or with cells of the sorogen that encysted without developing into spores. The spores are easily sloughed off and may not have been present in the sections studied.

B. *Copromyxa arborescens*

The fine structure studies by Nesom (1973) on this species further confirm the similarities between sorocysts and microcysts. Autophagic

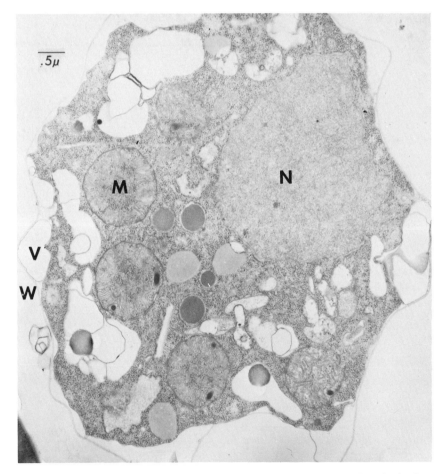

Fig. 207. *Guttulinopsis vulgaris,* ultrastructure of spore. Note nucleus (N), clear vesicles (V) expelled between protoplast and thin spore wall (W), and mitochondria (M) with disc-shaped cristae. Micrograph by M. Dykstra.

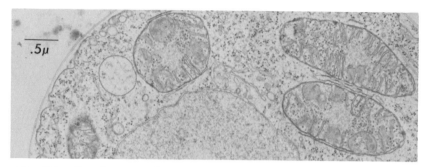

Fig. 208. *Guttulinopsis vulgaris,* ultrastructure of spore. Note thin spore wall and mitochondria with disc-shaped cristae. Micrograph by M. Dykstra.

vacuoles are present in both. The wall is composed of a single relatively thick fibrillar layer (Fig. 205). The nucleus, especially during wall development, is characterized by a remarkable irregularity in the outline of its envelope (Figs. 205 and 206). This irregularity results from numerous bleblike, ribosome-coated evaginations, the function of which is uncertain. Golgi bodies are prominent in the cells during sorogenesis. Thin sections through the sorocarp show that the columns of sorocysts are surrounded by a thin membranous layer.

The mitochondria of *C. arborescens* are unlike those of *Acrasis* in that the cristae are tubular rather than platelike. In addition, evaginations of the outer mitochondrial membrane enclosing electron-translucent contents develop during aggregation, much as in the prespore cells of *Dictyostelium discoideum.* Their function in *Copromyxa* has not been determined.

C. Guttulinopsis vulgaris

There have been no intensive ultrastructural studies of this species. In a preliminary examination of amoebae and spores, Mr. Michael Dykstra, a graduate student in our laboratory, has found that the mitochondria contain platelike cristae and the spore wall is very thin and irregular in outline (Figs. 207 and 208). The withdrawal of the protoplast here and there from the spore wall results at least in part from the expulsion of large vacuoles from the cytoplasm.

Although still of a limited nature, developmental and ultrastructural studies of acrasid cellular slime molds clearly indicate that these organisms are far from being a homogeneous group. Future reports on other members of the class are awaited with much interest for the light that they may shed on this subject.

Associated Groups

Plasmodiophorina

and

Labyrinthulina

CHAPTER 6

Plasmodiophorina (Plasmodiophorids)

I. INTRODUCTION

The plasmodiophorids are a group of obligate, endobiotic parasites that have been variously classified among the fungi, mycetozoans, Proteomyxa, and Rhizopoda. The most thorough treatment of the group is that of Karling (1968), whose monograph includes exhaustive discussions of their taxonomy, distribution, parasitism, and phylogeny. Only two species, one causing club root of cabbage and other cruciferous plants and the other causing powdery scab of the white potato, are of any noteworthy economic importance. The fact that so many investigators have studied these organisms is probably related less to their importance as plant pathogens than to the intriguing, unanswered questions concerning their life cycles and their relationships to other groups. In a discussion of the possible phylogenetic relationships of plasmodiophorids, Karling (1942, 1968) has favored the concept that they had an independent origin from the Proteomyxa, a group generally classified among the Euprotista. This viewpoint is tentatively adopted here.

The Plasmodiophorina are included in this text simply because they have been so frequently allied with the Mycetozoa, especially the myxomycetes, and because their true relationships remain uncertain. As in the slime molds the major trophic stage is a naked multinucleate protoplast that may be considered a plasmodium, and the motile stage is an anteriorly biflagellate zoospore with nonmastigonemate flagella. As will become apparent, however, there are few other similarities and the differences between the two groups indicate a lack of any close relationship.

II. OCCURRENCE IN NATURE AND LABORATORY MAINTENANCE

The plasmodiophorids appear to be fairly generally distributed throughout the world, although some species are much less commonly reported than others. Karling recognizes a single order with one family, nine genera, and approximately 35 species. He notes that five of these species occur in filamentous fungi, four in algae, one in pteridophytes, and the remainder in cells of higher plants. The two economically important species are *Plasmodiophora brassicae* Wor., which causes club root of cruciferous plants, and *Spongospora subterranea* (Wallr.) Lagerh., which causes powdery scab of the white potato. Both have become fairly worldwide in distribution.

Certain plasmodiophorids are known to parasitize only a single host species, while others attack a wide variety of them. In the genus *Ligniera*, for example, *L. betae* (Nemec) Karling is known only as a parasite in the roots of the common beet, whereas *L. junci* (Schwartz) Maire and Tison attacks the root hairs and roots of a great variety of monocotyledonous and dicotyledonous plants. Whether a species such as *L. junci* is actually comprised of numerous varieties, or at least pathogenic races, can be answered only by additional studies that include cross-inoculations of host plants.

The greatest block to detailed investigations of plasmodiophorid life cycles has been their obligate parasitism. Most studies have been performed on naturally infected host plants or on plants artificially infected by the investigator. This has occasionally led to the inclusion of contaminating organisms in the resultant reports. An improvement on these methods has recently been devised by Ingram (1969) and Williams *et al.* (1969), who have been able to maintain *Plasmodiophora brassicae* indefinitely in cabbage callus tissue grown on synthetic media. Young infected callus tissues grow more rapidly than uninfected ones. The technique should prove very useful for investigating further the life cycles of this and other species of the group, especially since all stages of development from minute plasmodia to mature cysts, and occasionally even zoosporangia, occur within the callus cells and the mature cysts are germinable and infective.

III. LIFE CYCLE

In our present state of knowledge about plasmodiophorids, any attempt at a complete description of the life cycle is hazardous. Certain aspects remain poorly understood and the subject of conflicting interpretations. Nevertheless, if previous reports are viewed in the light of recent investigations, a better concept of the life cycle begins to emerge. The tentative life cycle diagrammed in Fig. 209 is a composite one in which the position of certain events, especially plasmogamy, karyogamy, and meiosis, need additional confirmation. Although most of the information in the diagram is derived from reports on the more frequently studied species *P. brassicae, Spongospora subterranea,* and *Sorosphaera veronicae,* it

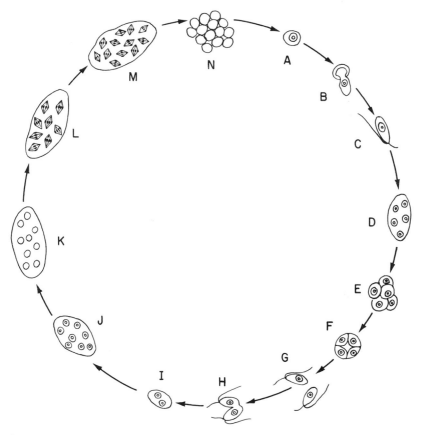

Fig. 209. Diagram of the plasmodiophorid life cycle (tentative): (A) cyst; (B) germinating cyst; (C) primary zoospore; (D) primary plasmodium; (E) sporangiosorus; (F) sporangium with secondary zoospores; (G) secondary zoospores; (H) secondary zoospores functioning as gametes; (I) binucleate product of plasmogamy (young secondary plasmodium); (J) multinucleate secondary plasmodium; (K) meiotic prophase ("akaryote" stage); (L) first meiotic division; (M) second meiotic division; (N) cysts. Karyogamy probably occurs just before meiotic prophase. See text for possible variations.

is supplemented by discoveries on several others. Karling (1968) has discussed the many variations in the life cycle of these organisms postulated by other investigators.

When the uninucleate cyst, or resting spore, germinates a single primary zoospore emerges. It has two apical, unequal, smooth flagella, the shorter one directed anteriorly and the longer one, tapered at the end, directed posteriorly during swimming (Ledingham, 1935). Although a few investigators have believed that plasmogamy occurs between primary zoospores, most have not found support for this viewpoint. It is generally thought that primary zoospores infect the host organism and produce the primary plasmodia.

Within the primary plasmodium a type of intranuclear division referred to as "promitotic" or "cruciform" has been described, the latter term being preferred because a true spindle is involved in the separation of chromosomes and the division is therefore mitotic in nature. The unique feature of the cruciform division, which has also been reported in certain euprotists, is the elongation and division of the nucleolus during the constriction of the nucleus into two daughter nuclei. At metaphase a ring of chromosomes encircles the nucleolus and, in side view, the ring and nucleolus give the mitotic figure a cruciform appearance. A ring of daughter chromosomes passes to each end of the nucleus along with a daughter nucleolus. This is followed by constriction of the persistent nuclear membrane around the middle to form the two daughter nuclei. A somewhat different version has been expressed by Keskin (1964), who reported that in *Polymyxa betae* metaphase has a cruciform appearance but the nucleolus degenerates before passage of the two chromosomal rings to opposite poles. Differential staining techniques (e.g., Feulgen) have demonstrated that the nucleolus is probably homologous with that of higher organisms.

The mature primary plasmodium which may be relatively small, cleaves into uninucleate segments that comprise the sporangiosorus. Typical mitotic divisions have been reported in the developing zoosporangia (Köle, 1954; Miller, 1958), and four to many secondary zoospores, similar in appearance to the primary zoospores, are produced and make their exit through an opening in the thin sporangial wall or through a discharge tube (Figs. 210–212).

Several investigators have reported that the secondary zoospores fuse in pairs to produce the secondary plasmodia (Cook, 1933; Köle, 1954; Keskin, 1964; Tommerup and Ingram, 1971). However, Köle disputed Cook's contention that karyogamy follows in the fusion product, claiming instead that the two nuclei remain separate and that the fusions are of a nonsexual nature. It has also been claimed that amoeboid protoplasts

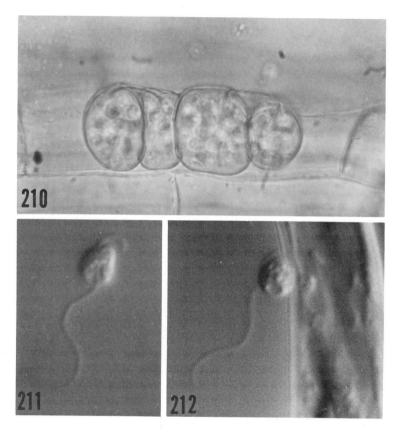

Figs. 210–212. Zoospores of plasmodiophorids. Fig. 210. *Sorosphaera veronicae,* zoosporangia in root cells discharging zoospores, ×816. Fig. 211. *Plasmodiophora brassicae,* primary zoospore with short anterior and long posterior flagellum. Phase contrast ×3000. Fig. 212. Same, zoospore attached to cabbage root hair. Phase contrast ×3000. Fig. 210 from Miller (1958); Figs. 211 and 212 from Aist and Williams (1971).

may fuse within the host cells (Tommerup and Ingram, 1971). Still other reports state that unfused secondary zoospores may cause infection and produce secondary plasmodia.

As the secondary plasmodium enlarges within the host cell, its nuclei undergo synchronous intranuclear cruciform divisions (Figs. 213 and 214). The most commonly reported number of chromosomes in the chromosome ring is four. The plasmodium may also undergo plasmotomy (schizogeny) within the host cell. As the plasmodium approaches maturity, the nuclei reach a stage at which they become indistinct (the "akaryote" stage), probably as a result of degeneration of the nucleoli. In

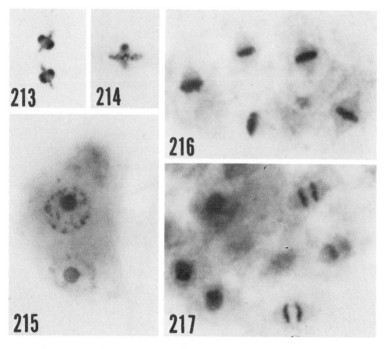

Figs. 213–217. *Sorosphaera veronicae;* cytology. Figs. 213 and 214. Cruciform divisions. Note ring of chromosomes around dividing nucleolus in each nucleus. ×2425. Fig. 215. Interphase nucleus in zoosporangial plasmodium, ×2425. Fig. 216. Metaphase of first sporogenic (meiotic?) division, ×2910. Fig. 217. Anaphase of same, ×2910. From Miller (1958).

light of recent investigations, it now seems likely that this is prophase of the upcoming nuclear division, which is typically mitotic in appearance (Fig. 215). It is generally believed that there are two successive nuclear divisions at this time in which the chromosomes separate on spindles and the nucleolus and nuclear membrane are absent (Figs. 216 and 217). A number of investigators have claimed that these two divisions are meiotic, and there are now several indications that this view is correct. Tommerup and Ingram (1971), studying the life cycle of *P. brassicae* in tissue cultures, found that the secondary zoospores fuse in pairs, the fusion product becoming a secondary plasmodium without an intervening karyogamy. Their study of Giemsa-stained preparations led them to conclude that karyogamy immediately precedes the two terminal nuclear divisions in the plasmodium, which they claim are meiotic. This agrees with Osborne's report (1911) describing stages in nuclear fusion followed by meiosis just before spore formation in

Spongospora subterranea. Whiffen (1939) has reported similar developments in *Octomyxa achlyae* Couch, with karyogamy occurring just prior to meiosis, during which quadripolar meiotic spindles develop. Perhaps

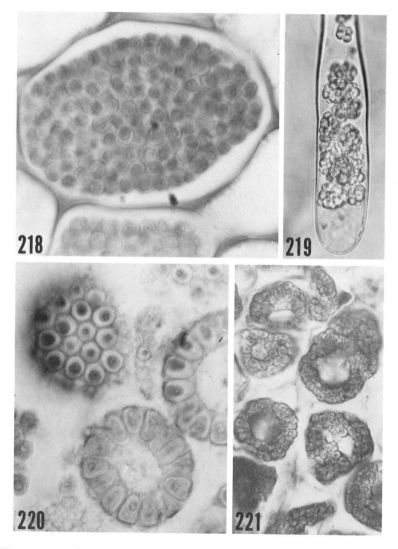

Figs. 218–221. Cysts of plasmodiophorids. Fig. 218. *Plasmodiophora brassicae,* loosely clustered cysts in cabbage root cell, ×643. Fig. 219. *Ligniera junci,* small cystosori of variable size in root hair of chickweed, ×643. Fig. 220. *Sorosphaera veronicae,* maturing hollow cystosori in root cell of *Veronica,* ×1455. From Miller (1958). Fig. 221. *Spongospora subterranea,* cystosori in cells of potato tuber, ×200.

the most convincing evidence that meiosis occurs in the plasmodium is found in the brief report by Braselton and Miller (1973) of synaptonemal complexes in meiotic prophase I nuclei (Fig. 231) of *Sorosphaera veronicae* Schroeter.

Following these two divisions the plasmodium cleaves into uninucleate cysts that occur either loosely in the host cell (Fig. 218) or in adherent small groups (Fig. 219), hollow or spongy spheres (Figs. 220 and 221), or flattened two-layered discs. In *S. veronicae*, Miller (1958) reports that the first nuclear division is followed by cleavage of the plasmodium into uninucleate segments. Following the second division these cleave into two incipient spores each. The cysts of *P. brassicae* may retain their viability in the soil for as long as 8 years. The cyst walls of plasmodiophorids appear to lack cellulose.

It is generally conceded that plasmodiophorids have a primarily osmotrophic form of nutrition. However, several exceptions have been noted. Engulfment of solid particles by amoeboid zoospores of *Polymyxa graminis* Ledingham (1933) and of algal cells by young plasmodia of *Ligniera junci* (Maire and Tison, 1911) has been described, and several publications, including a recent ultrastructural study (Williams and McNabola, 1967), have presented evidence that more mature plasmodia may ingest portions of the host protoplast.

There have been several reports (Rao and Brakke, 1969; Kusaba and Toyama, 1970; Toyama and Kusaba, 1970) that the zoospores of *Polymyxa graminis* may serve as vectors of wheat and barley mosaic virus.

IV. CLASSIFICATION

The many and varied taxonomic treatments of the plasmodiophorids that have appeared in the literature since their discovery have been discussed in detail by Karling (1968). Some idea of the variety of viewpoints can be obtained from a few examples. Zopf (1885) established the family Plasmodiophoraceae within his zoosporic group Monadineae, which he included in the myxomycetes. Lankester (1885), adopting a new family name for the plasmodiophorids, transferred the Monadineae to his new class Proteomyxa of the Protozoa. Schroeter (1897), also ignoring Zopf's family name, created a new order, family, and class (Phytomyxinae), which he placed near the myxomycetes. Cook (1928, 1933) accepted the view that plasmodiophorids were not related to the fungi but had a separate origin from the proteomyxans. He also proposed that no organism lacking cruciform divisions should be placed in the group. Kudo (1963) treats both the Proteomyxa (containing the plasmodiophorids) and Mycetozoa as orders of the subclass Rhizopoda, class Sarcodina. In his taxonomic treatment of the fungi, Bessey (1950) included the order Plasmodiophorales in the subclass Mycetozoa, which he felt belonged in the class Sarcodina of the Protozoa. The latest treatment by Sparrow (1960) recognizes the group as the first and lowest order of the biflagellate series of the phycomycetes but considers them of independent origin. Alexopoulos (1962), using the class name Plasmodiophoromycetes and the common name "endoparasitic slime molds," discusses them at the end of the biflagellate series of lower fungi, although he accepts Sparrow's viewpoint that they have probably arisen independently of other groups of fungi. A committee of the Society of Protozoologists (Honigberg et al., 1964) has recognized the plasmodiophorids as an order of the subclass Mycetozoia, class Rhizopodea, and superclass Sarcodina of the Protozoa. Finally, the group has most recently been treated as class Plasmodiophoromycetes in subdivision Mastigomycotina of the division Eumycota (Waterhouse, 1973).

The variety of taxonomic treatments largely reflects the lack of sufficient information on the life cycles and biology of the group. The presence of plasmodia and differences between their zoospores and those of commonly recognized groups of aquatic fungi, among other factors, sharply distinguish the plasmodiophorids from organisms generally considered as fungi. The type of life cycle that is now becoming apparent, the occurrence of cruciform divisions in the plasmodia, the obligate parasitism, and the absence of true fruiting bodies in plasmodiophorids also clearly distinguish them from the Mycetozoa. In view of certain indications (to be discussed below) that the group may have originated independently of either fungi or mycetozoans, but possibly from the Proteomyxa of the Euprotista, the plasmodiophorids are not considered closely related to the fungi or the Mycetozoa.

A second group of recently discovered endobiotic, plasmodial parasites of ciliates, the Endemosarcidae (Erdos and Olive, 1971), tentatively allied with the Proteomyxa, has some of the characteristics of plasmodiophorids but apparently lacks others, such as cruciform nuclear division. The single known genus *Endemosarca*, with three species, is discussed below along with *Phagomyxa algarum* Karling (1944), an endobiotic species in cells of marine algae that has characteristics somewhat intermediate between those of plasmodiophorids and proteomyxans.

In the following treatment a single class, the Plasmodiophorea, of the subphylum Plasmodiophorina is recognized. Thereafter, the system of Karling is followed in the recognition of a single order (Plasmodiophorida) and family (Plasmodiophoridae) containing nine genera and approximately 35 species. The key is based primarily on the manner in which the spore cysts are arranged—whether in loose masses or closely aggregated into cystosori of various types. As Karling and others have noted, there are confusing intergradations among genera and species with respect to this character and there is no question that further studies are needed to define more clearly the generic and specific boundaries.

Key to the Genera of Plasmodiophoridae
(adapted from Karling, 1968)

1. Cysts occurring in loose masses......................... *Plasmodiophora*
1. Cysts united into cystosori
  2. Cystosori more or less definite in shape and size
    3. Cystosori small (mostly 2- to 8-spored)
      4. Cystosori mostly 2- to 4-spored..................... *Tetramyxa*
      4. Cystosori 8-spored............................... *Octomyxa*
    3. Cystosori larger, multispored
      5. Cystosori spherical to ellipsoidal, hollow or spongy within
        6. Cystosori hollow spheres or ellipsoids .... *Sorosphaera*
        6. Cystosori spongy balls of spores .... *Spongospora*
      5. Cystosori mostly flattened, two-layered disks........... *Sorodiscus*
  2. Cystosori indefinite in shape and size
    7. Zoosporangia small and numerous, united into loose sporangiosori .... *Ligniera*
    7. Zoosporangia larger, occurring singly or in sporangiosori
      8. Zoosporangia not in distinct sporangiosori, lobed, irregular, with elongate exit tubes................... *Polymyxa*
      8. Zoosporangia occurring singly or in loose sporangiosori; spherical, with exit papilla or exit tube .... *Woronina*

The earlier concept of Lankester (1885) that the plasmodiophorids are related to proteomyxans has recently been revived by several investigators (Karling, 1968; Olive, 1970; Erdos and Olive, 1971). However, the Proteomyxa are a heterogeneous and poorly known group and meaningful comparisons can be made only with certain members of the order, such as *Aphelidiopsis, Pseudosporopsis,* and *Gymnococcus,* and even these are incompletely understood. *Aphelidiopsis epithemiae* Scherffel (1925), which occurs in diatoms and ingests portions of the host's protoplasts, develops plasmodium-like amoeboid protoplasts that are undoubtedly multinucleate, although the cytology has not yet been studied. The plasmodium eventually cleaves into two to eight segments that develop thin surrounding walls and become zoosporangia (zoocysts) in which zoospores are delimited. The zoospores are released by deliquescence of the sporangial wall. Resting spores (probably resting sporangia) are also produced. The zoospores appear similar to those of plasmodiophorids and swim with the shorter flagellum forward and the longer one behind. Zopf (1885) based his description of *Gymnococcus* on three species found in cells of algae. They were described as parasites, and phagotrophy was not mentioned. Both zoosporangia and resting spores are produced. The zoospores are biflagellate. *Gymnococcus cladophorae* de Bruyne (1890), found in the cells of the green marine alga *Cladophora*, has been described as phagotrophic. It is doubtful whether species and genera with uniflagellate zoospores that have been placed in the Gymnococcaceae should be maintained in the family. In addition, detailed cytological studies with light and electron microscopes are needed to clarify various aspects of the gymnococcid life cycle, including whether meiosis and cruciform mitosis occur.

Phagomyxa algarum Karling (1944), a phagotrophic plasmodial organism occurring in the cells of marine algae, forms sporangiosori and zoospores similar in appearance and motility to those of plasmodiophorids, but resting spores are unknown. Synchronous mitoses were observed in the plasmodia, five or six chromosomes being reported on a spindle with centrosome-like structures at the poles, but cruciform divisions were not demonstrated. Karling considered the organism a possible transitional form between proteomyxans and plasmodiophorids.

Of more recent description is the family Endemosarcidae Olive and Erdos, with the single genus *Endemosarca* Olive and Erdos containing three known species parasitic within ciliates (*Colpoda* spp.): *E. hypsalyxis* Erdos and Olive (1971) (Fig. 222), *E. ubatubensis* Erdos and Olive (1971) (Fig. 223), and *E. anomala* Erdos (1973). They are recovered with their hosts from soil, humus, rotting wood, etc., that have been plated out on agar media of the types used in isolating protostelids (Chapter 2). All are characterized by zoospores similar in general appearance to those of plasmodiophorids (Fig. 224). The longer, posteriorly oriented flagellum is directed along a longitudinal groove in the zoospore. Both flagella are nonmastigonemate (Fig. 225). In *Endemosarca*, as in plasmodiophorids and *Phagomyxa*, when the zoospore slows down, the anteriorly directed flagellum begins to beat with a laterally directed stroke. The zoospore penetrates the ciliate cell directly and enlarges within the host protoplasm to form a small plasmodium (Fig. 226) that becomes multinucleate. The nuclear envelope apparently persists during mitosis but no evidence of cruciform division has been found (Erdos, 1972a). A chromosome number of five or six was observed.

The most remarkable feature of *Endemosarca* is the development within the maturing plasmodium of an exit tube, which is long and coiled in *E. hypsalyxis*, shorter and broader in *E. ubatubensis*, and absent in *E. anomala*. The exit tube is continuous with the thin sporangial wall. Another interesting feature of the genus relates to the plasmodium, which late in its development engulfs a major portion of the host proto-

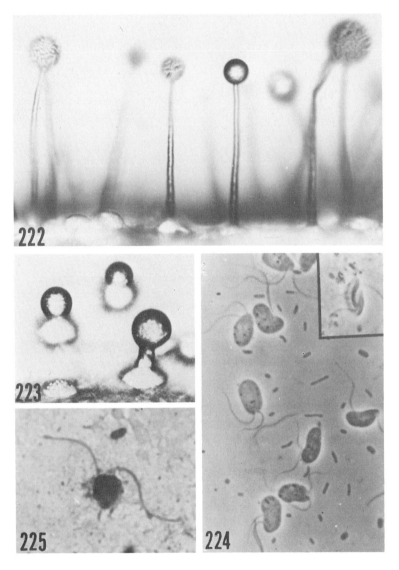

Figs. 222–225. *Endemosarca*. Fig. 222. "Fruiting bodies" of *E. hypsalyxis* on agar surface, ×1000. Fig. 223. "Fruiting bodies" of *E. ubatubensis* on agar surface, ×550. Fig. 224. Anteriorly biflagellate zoospores of *E. ubatubensis*. Inset shows longitudinal groove, ×1575. Fig. 225. Zoospore of *E. hypsalyxis* treated with Loeffler's stain to show the two nonmastigonemate flagella, ×2360. Figs. 222, 223, and 225 from Erdos and Olive (1971); Fig. 224 from Erdos (1971).

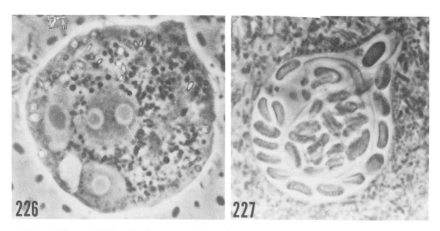

Figs. 226 and 227. *Endemosarca hypsalyxis.* Fig. 226. Cell of parasitized ciliate with young uninucleate and binucleate plasmodia (host nucleus at left), ×2380. Fig. 227. Sporangium prior to discharge, showing encysted zoospores and coiled discharge tube inside, ×2125. From Erdos and Olive (1971).

plasm. The engulfed protoplasm thereafter appears as a large, somewhat eccentric food vacuole within the plasmodium (Erdos, 1972b). A thin sporangial wall develops around the entire plasmodium and uninucleate zoospores are delimited. In *E. hypsalyxis* the exit tube can be seen coiled among the spores (Fig. 227). Thick-walled resting spores and sexual reproduction are unknown.

Before they encyst within the zoosporangium, the zoospores develop flagella and are very active for a brief period. By this time the host cell is dead. If the infected cells are lying on a substrate, such as agar, with little or no free water on the surface, the exit tubes evaginate rapidly, vertically, and aerially, breaking open at the tip and discharging their glutinous masses of elliptical to somewhat bean-shaped spores. The spore masses appear ideally suited to being picked up and dispersed by small insects and mites. The empty sporangium remains as a dome-shaped base (Figs. 222 and 223). In *E. hypsalyxis* the entire process of exit tube evagination and spore discharge may take place within 2 minutes. The complete life cycle requires only 2 days.

When encysted spores are placed in a droplet of water they fail to germinate, but if a few host cells are added they germinate readily, each spore giving rise to a single zoospore through a slit in the thin spore wall.

Endemosarca hypsalyxis is a common species that has been isolated from various parts of the world. *Endemosarca ubatubensis* has been collected twice in Brazil and once in North Carolina. *Endemosarca anomala* has also been found twice, once on a collection of old soursop fruits in Cousin Island, Seychelles and again on dead plants of squaw root in Highlands, North Carolina.

Arnaud (1948, 1949) described a new family of six genera that he called the Heimerliaceae, a name since shown to be nonvalid (Erdos and Olive, 1971). These organisms were found in France in decaying vegetation on the ground during rainy periods. All appeared to have globular masses of spores borne aerially at the tips of slender stalklike structures, but the spores were found to be nongerminable and no developmental details on the group were recorded. Although Arnaud considered them relatives of dictyostelids, this appears unlikely. It is quite possible that some, if not all, are parasites of protozoans and related to *Endemosarca*. If future studies prove this to be the case, they should probably be placed in the Endemosarcidae.

When further information becomes available on these peripheral groups, it may be found advisable to recognize a second order within the class Plasmodiophorea to accommodate forms resembling plasmodiophorids but lacking cruciform mitosis or other features considered essential to the integrity of that group. Such an order might contain the Gymnococcidae (emended) of the Proteomyxa, *Phagomyxa*, *Endemosarca*, and possibly Arnaud's genera.

V. ULTRASTRUCTURE

A. Plasmodiophorids

Fine structure studies of *Plasmodiophora brassicae* (Williams and Yukawa, 1967; Williams and McNabola, 1967) reveal that the multinucleate plasmodium is enclosed by the host cytoplasm, the latter showing an enrichment of organelles and containing an enlarged nucleus. Within the plasmodium are found abundant rough ER and free ribosomes, numerous lipid bodies, Golgi bodies, uninucleolate nuclei associated with cytoplasmic microtubules, and abundant mitochondria with relatively few tubular cristae. The plasmodium was first reported to be bounded by two unit membranes, but later Williams and McNabola (1970) concluded that the outer membrane arose by invagination of the host plasmalemma during initiation of infection. Convincing evidence of this is lacking. Prior to sporulation the outer membrane disappears and nuclear divisions among several plasmodia in a host cell, which up until then have been synchronous within individual plasmodia, become synchronous among all the plasmodia in the cell. Fine structure studies on *Polymyxa*

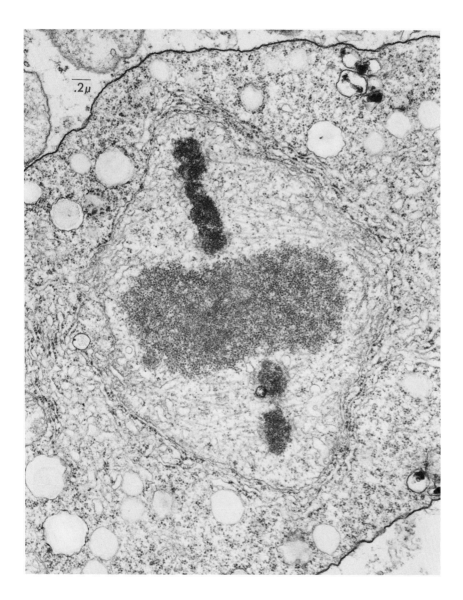

Fig. 228. *Sorosphaera veronicae;* electron micrograph of cruciform division. Note nucleus surrounded by abundant rough ER, dividing nucleolus, equatorial plate of chromosomes, spindle microtubules, and persistent nuclear membranes. Courtesy of J. P. Braselton and C. E. Miller, Department of Botany, Ohio University.

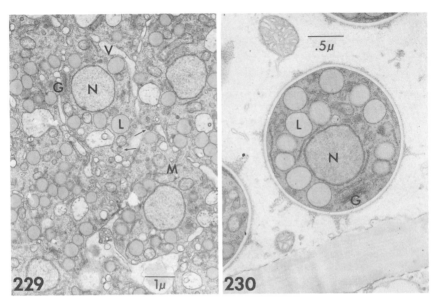

Figs. 229 and 230. *Plasmodiophora brassicae;* ultrastructure of cyst development. Fig. 229. Cleavage stage in plasmodium, showing alignment and coalescence of cleavage vacuoles leading to cyst formation. Fig. 230. Mature cyst with spinelike structures on the wall. G, Golgi apparatus; L, lipid body; M, mitochondrion; N, nucleus; V, cleavage vacuoles; arrows denote small vesicles with granular matrix. From Williams and McNabola (1967).

betae (Keskin, 1971) and *Sorosphaera veronicae* (J. P. Braselton and C. E. Miller, unpublished) confirm that somatic nuclear divisions are of the cruciform type (Fig. 228). Williams and McNabola have described invaginations of the plasmalemma of maturing plasmodia involving engulfment of host cytoplasm containing such organelles as plastids and mitochondria, but serial sections must be examined to determine whether the invaginations result in the formation of complete food vacuoles. After breakdown of the outer membrane, organization of the host protoplast is disrupted, indicating that this may have been a host membrane that functioned as a defense mechanism against the parasite.

The beginning of cyst formation is detectable by the appearance in the cytoplasm of numerous vacuoles that become aligned around uninucleate portions of the cytoplasm (Fig. 229). These vacuoles fuse together and the resultant inner membrane forms the plasmalemma of the incipient cyst. Granular material from the intercyst matrix is deposited on the surface of the developing cyst in the form of small spines, after which the cyst wall develops between plasmalemma and spines (Fig. 230). Throughout cystogenesis small membrane-bounded vesicles containing a granular matrix are found in the cytoplasm (Fig. 229). It will be interesting to learn whether these represent a kind of prespore vesicle that contributes to cyst wall formation.

Only in *Sorosphaera veronicae* has ultrastructural evidence of meiosis

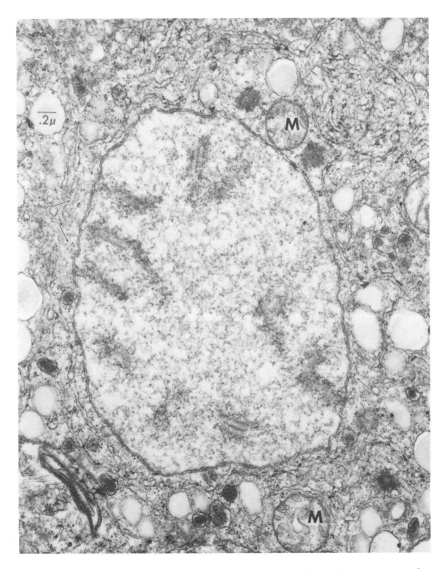

Fig. 231. *Sorosphaera veronicae;* electron micrograph of meiotic prophase nucleus containing synaptonemal complexes. Note microtubules (arrows) just outside nucleus; mitochondria (M). Courtesy of J. P. Braselton and C. E. Miller.

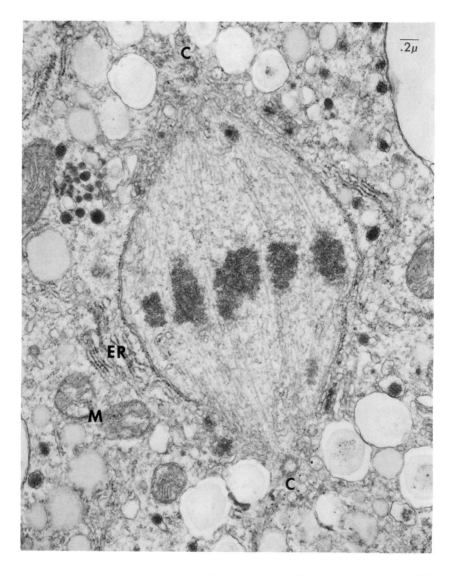

Fig. 232. *Sorosphaera veronicae;* electron micrograph of metaphase, probably of first meiotic division. Note well-developed spindle of microtubules, metaphase plate of chromosomes, persistent nuclear envelope with polar openings, centrioles (C), mitochondria (M), and rough endoplasmic reticulum (ER). Courtesy of J. P. Braselton and C. E. Miller.

been reported. Braselton and Miller (1973) illustrated synaptonemal complexes in what appear to be meiotic prophase nuclei in the plasmodia (Fig. 231). They also demonstrated the presence of centrioles at the poles of nuclei thought to be undergoing the first meiotic division (Fig. 232). The nuclear membrane remains intact except at the poles, where it is disrupted by the spindle microtubules.

Examinations of zoospores under the electron microscope (Fig. 233) have clearly demonstrated that both flagella lack mastigonemes and that the posteriorly directed flagellum, which is about 3–3.5 times the length of the shorter, anterior one, has a tapered end piece (Köle and Gielink, 1961; Keskin, 1964).

A most unusual infection apparatus has been found within the encysted zoospores of *Polymyxa betae* (Keskin and Fuchs, 1969) and *Plasmodiophora brassicae* (Aist and Williams, 1971), indicating a high degree of specialization in these cells. About 2 hours after the zoospore has become attached to a root hair, a tubular structure (*Rohr*) containing an osmiophilic bulletlike spine (*Stachel*) develops within the encysted cell (Fig. 234). The spine is surrounded by a fibrillar adhesive material. Infection occurs when the tube rapidly evaginates; the adhesive matrix

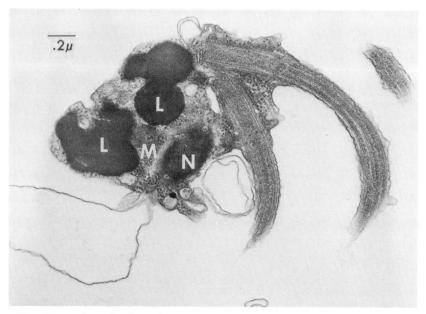

Fig. 233. *Plasmodiophora brassicae;* electron micrograph of primary zoospore. Note the pair of nonmastigonemate flagella, lipid bodies (L), mitochondrion (M), and nucleus (N). From Aist and Williams (1971).

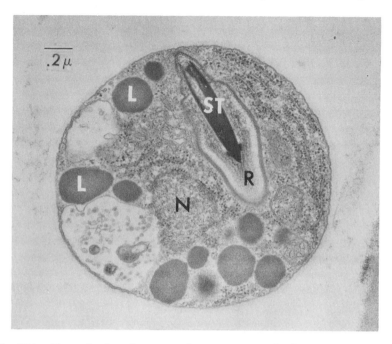

Fig. 234. *Plasmodiophora brassicae;* electron micrograph of an encysted primary zoospore before infection of host. The *Stachel* (ST) is enclosed in the *Rhor* (R), the limiting membrane of which is continuous with the plasmalemma. Note also the nucleus (N) and lipid bodies (L). From Aist and Williams (1971).

forms a kind of adhesorium on the host surface as the spine punctures the host cell wall and is followed about 1 second later by the parasite protoplast. The total time for adhesorium formation and host penetration is estimated to be about 1 minute (Aist and Williams, 1971). This method of infection is characteristic of both primary and secondary zoospores. It will be of interest to learn whether this unique and remarkable mode of infection is characteristic of plasmodiophorids in general.

B. Endemosarca

Fine structure studies of *E. hypsalyxis* and *E. ubatubensis* have revealed a similar pattern of development in both (Erdos, 1972b). Shortly after infection the young plasmodium becomes surrounded by vesicles of host rough ER. These are at first oriented perpendicularly to the plasmodial plasmalemma (Fig. 235) but soon become parallel to it and fuse together, forming two separate membranes around the plasmodium. The inner host ER membrane loses its ribosomes and becomes closely applied to the plasmodial plasmalemma. Shortly thereafter, narrow invaginations of the parasite plasmalemma and inner host ER membrane occur (Fig. 236). Some of these invaginations may be precursors of the internally formed discharge tube.

The plasmodial cytoplasm contains many ribosomes, rough ER cisternae, lipid bodies, mitochondria, and uninucleolate nuclei. The mitochondria are often irregular or branched and contain relatively few tubular cristae, in some areas none, and they therefore resemble the mitochondria of certain other parasitic protists. No typical Golgi bodies were found, but nuclear-derived amplexi thought to represent rudimentary dictyosomes are present. Although few details of nuclear division were observed, one late telophase figure and observations under the phase microscope indicate that the nuclear envelope persists and the nucleolus degenerates during somatic mitosis. Centriole-containing plaques located in pockets of the nuclear envelope were found associated with microtubules directed toward the nuclear envelope.

Just prior to spore cleavage in *E. hypsalyxis* the nuclei become periph-

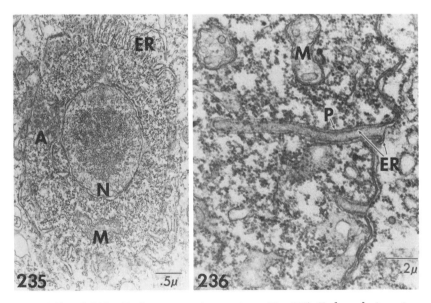

Figs. 235 and 236. *Endemosarca;* ultrastructure. Fig. 235. *E. hypsalyxis,* uninucleate plasmodium showing early stage in development of envelope of host endoplasmic reticulum (ER). Note the nucleus (N) with central nucleolus, the amplexus (A) derived from outer nuclear membranes, and mitochondrion (M). Fig. 236. *E. ubatubensis,* plasmodial invaginations consisting of the plasmodial plasmalemma (P) and inner membrane of the host endoplasmic reticulum envelope (ER). From Erdos, 1972b.

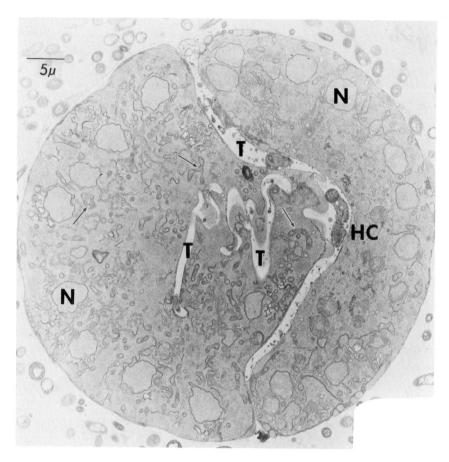

Fig. 237. *Endemosarca hypsalyxis;* plasmodium developing into sporangium, showing sinuous discharge tube (T), peripheral nuclei (N), and irregular, often branched mitochondria (arrows). Note area of engulfed host cytoplasm (HC). From Erdos (1972b).

erally aligned and the sinuous discharge tube develops within the plasmodium (Fig. 237). The thin sporangial wall is laid down at about the same time between the plasmodial plasma membrane and the envelope of host ER. Cytoplasmic vesicles that fuse with the plasma membrane probably contribute to wall formation. The spores are delimited by means of cleavage vesicles.

Further life history and ultrastructural studies of plasmodiophorids and possibly related groups are needed before a better understanding of the true relationships of the subphylum can be gained.

CHAPTER 7

Labyrinthulina
(Labyrinthulas and Thraustochytrids)

I. INTRODUCTION

The subphylum Labyrinthulina is recognized here as including a single class with one order and two families, the Labyrinthulidae and the Thraustochytriidae. The first-mentioned family was erected by Haeckel (1868) to accommodate Cienkowsky's genus *Labyrinthula,* which was based on the description of two species, *L. vitellina* Cienkowsky and *L. macrocystis* Cienkowsky (1867), found growing on marine algae in the Black Sea near Odessa, Russia. Subsequent treatments have placed the group among the Mycetozoa, the Fungi, and in the Rhizopodea (both in the Amoebina and the Proteomyxida) of the Protozoa. Much of the uncertainty about their proper taxonomic position has resulted from insufficient knowledge of their life cycles and structure. This paucity of information has resulted in the description of several genera, including *Chlamydomyxa* Archer (1875) and *Pseudoplasmodium* Molisch (1926), that are now considered synonymous with *Labyrinthula.* In addition, the recently described genus *Labyrinthodictyon* Valkanov (1972) does not appear to differ significantly from *Labyrinthula* and should probably be reduced to synonymy with it. The description of *Labyrinthomyxa* Dubosq (1921) is almost certainly based on a mixture of at least two different organisms, including a labyrinthulan. We therefore do not consider the taxon valid. The description of *Labyrinthorhiza* Chadefaud (1956) indicates that it is not a labyrinthulan at all but a protist with anastomosing cells. About seven or eight species of *Labyrinthula* are known.

Labyrinthulas are found primarily in marine habitats growing on a variety of algae or on leaves of marine angiosperms, such as eel grass. They are generally considered to be at least partly parasitic. Some are

215

involved in the "wasting disease" of eel grass, a problem of considerable importance to the ecology of littoral marine life. A single freshwater species occurring on *Vaucheria* is known.

The labyrinthulas are particularly characterized by their spindle-shaped cells, which produce elongate hyaline extensions that eventually form a network. The latter has been variously termed "net plasmodium," "rhizoplasmodium," "filoplasmodium," "ectoplasmic net," etc. The spindle-shaped cells move within the net elements. For reasons discussed in Section V, the term "ectoplasmic net" (Perkins, 1972) is adopted here.

The thraustochytrids were first discovered by Sparrow, who described the genus *Thraustochytrium* (1936) and family Thraustochytriaceae (1943), later (1960) emending the family to include a second genus, *Japonochytrium* Kobayasi and Ookubo (1953). These organisms were described as being eucarpic and monocentric, with endobiotic rhizoidal system, resembling chytrids but producing planonts like those of oomycetous fungi, among which Sparrow was inclined to classify them. Eventually, two additional genera, *Schizochytrium* Goldstein and Belsky (1964) and *Aplanochytrium* Bahnweg and Sparrow (1972), were described. The publications of Perkins (1973a, 1974c) on *Labyrinthuloides*, in addition to unpublished data supplied by him, have convinced us that this genus should also be placed among the thraustochytrids. Recent chemical analysis of the cell walls of thraustochytrids, combined with ultrastructural studies, strongly indicate that these organisms are not oomycetous in nature but are related to *Labyrinthula*. For these reasons, we have decided to transfer them to the subphylum Labyrinthulina.

II. OCCURRENCE, ISOLATION, AND MAINTENANCE

The members of this subphylum are primarily marine organisms that are found growing on a wide variety of substrates, including green, brown, and red algae, as well as diatoms; on marine angiosperms, such as *Zostera* and *Spartina;* and on organic detritus. Detailed lists of substrates on which labyrinthulas have been found are given by Pokorny (1967, 1971). By baiting estuarine waters with a variety of natural substrates, including leaves and pollen of marine and terrestrial angiosperms, marine algae, yeasts, and tissues from crab and oyster, Perkins (1973a) was able to extend considerably the known substrates on which labyrinthulas and thraustochytrids can grow. He also demonstrated that these organisms are remarkably nonspecific with regard to substrate utilization. It is of interest to note that both groups are able to utilize animal tissues as a food source.

There has been much discussion about the degree of pathogenicity characteristic of the labyrinthulas. Although little information is available on development in nature, the ectoplasmic nets of some species are known to invade plant substrates, accompanied by the spindle-shaped cells. In *Zostera* deterioration of chloroplasts results. *Labyrinthula* species have been implicated repeatedly in the "wasting disease" of *Zostera marina* in Atlantic coastal waters of North America and Europe. The importance of eel grass to the ecology of littoral marine life and consequently to the fishing industry has evoked great interest in the role of labyrinthulas as pathogens. In reviewing this subject in considerable detail, Pokorny (1967) has noted the regular occurrence of *Labyrinthula* species in healthy beds of eel grass and has concluded that whereas these organisms are indeed parasitic on *Zostera*, the periodic devastations of eel grass are probably the result of a combination of factors, including changes in salinity, temperature, and quantity of sunlight during these periods.

Regardless of their degree of pathogenicity in nature, labyrinthulas are readily isolated and cultured in the laboratory. Pioneering studies on their culture and nutrition were pursued by Young (1937, 1943), Watson and Ordal (1951), Vishniac and Watson (1953), and Vishniac (1955a,b). Using the information published by these investigators, Pokorny (1971) devised a very satisfactory isolation medium containing 5% gelatin hydrolyzate, 5% horse serum, 20 mg% thiamine, 30 mg% streptomycin sulfate, 15 mg% penicillin G, 0.9% agar, and sea water, the finished medium being at pH 8. The antibiotics deter excessive growth of undesirable organisms, and the serum supplies sterol, a requirement of some (e.g., *L. vitellina*) but not all labyrinthulas. Several species (including *L. vitellina*) have also been shown to require exogenous thiamine. Carbon and nitrogen requirements may be satisfied by any of several amino acids as well as by gelatin hydrolyzate. When pieces of suspected host material (algae, *Zostera* leaves, etc.) are placed on such media, ectoplasmic nets with cells of the labyrinthula often advance onto the medium. In areas where the colony has outgrown contaminating organisms, blocks of agar containing portions of the colony may be cut out and transferred to fresh plates of medium. Following successive growth transfers on the same medium to insure elimination of contaminants, the organism may be maintained on the medium without antibiotics. Maintenance over a period of years is feasible if cultures are transferred on plates or in agar slants about every 3 weeks and stored at 5°C. Some species were grown and maintained on a more nearly defined medium by Vishniac and Watson (1953) and Vishniac (1955b). Recently, Sykes and Porter (1973) described the growth of an isolate resembling *L. vitellina* and found that it grew well axenically on a defined medium containing a combination of sodium glutamate, any of several carbon sources, thiamine, and inorganic nutrients.

Thraustochytrids are often first isolated by baiting collections of sea or estuarine water and sediments with pollen or other natural substrates, followed by transfer to suitable agar media. Media for axenic culture have been described by several investigators (Watson and Ordal, 1957; Goldstein, 1963a–c; Goldstein and Belsky, 1964; Perkins, 1973a). Although both labyrinthulas and thraustochytrids are able to utilize microorganisms such as bacteria, yeasts, and diatoms as a major food source in culture, they do not ingest these cells and digestion is osmotrophic (e.g., Klie and Mach, 1968). The axenic nutritional requirements of thraustochytrids resemble those of *Labyrinthula*.

There are large areas of the world in which no surveys of the Labyrinthulina have been undertaken. Undoubtedly, a greater diversity of genera and species will be reported when such studies have been made. Even more urgently needed at the moment, in addition to new information on life cycles, is a more standardized descriptive procedure. There is already too much confusion in the identification of isolates both at the generic and specific levels.

III. LIFE CYCLE

It is premature at this time to propose a representative life cycle for the Labyrinthulina. The only convincing evidence indicating that a sexual process may occur in these organisms is the demonstration of a typical meiotic prophase in a single unidentified isolate of *Labyrinthula* that resembles *L. algeriensis* (Perkins and Amon, 1969). It is interesting to note that the most detailed ultrastructural studies have been made on this and two other unidentified isolates of *Labyrinthula*, one studied by Pokorny (1971) and the other by Porter (1972). The following discussion of the labyrinthulan life cycle, insofar as it is understood, is therefore a composite one (Fig. 238). Certain developmental details are clarified further in the discussion of ultrastructure (Section V).

If a culture of *Labyrinthula* is initiated with a single spindle-shaped vegetative cell (Pokorny, 1971), the first observable development is the extrusion of several hyaline globules or sacs around the periphery of the cell. As these sacs enlarge, fine filamentous extensions develop from them (Fig. 240). Eventually the sacs anastomose, resulting in enclosure of the cell by inner and outer sac membranes, which in turn delimit

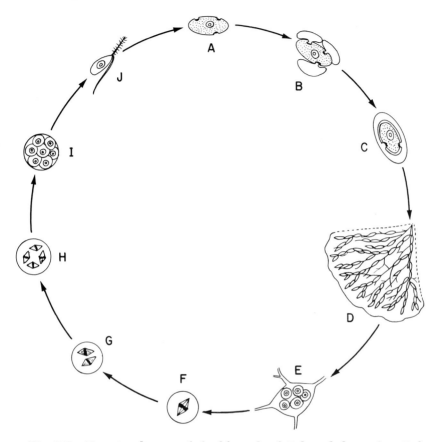

Fig. 238. Tentative diagram of the life cycle of *Labyrinthula* sp. (see Perkins and Amon, 1969): (A) spindle cell with sagenogens; (B) sagenogen producing ectoplasmic sacs; (C) spindle cell around which ectoplasmic sacs have fused; (D) developing colony; (E) cluster of zoosporangia in expanded portion of ectoplasmic net; (F–H) nuclear divisions in zoosporangium, probably meiotic; (I) mature zoosporangium with zoospores; (J) zoospore. Position of plasmogamy (if present) unknown but it is suspected of occurring between C and E.

a matrix that is often referred to as "slime" or "slime matrix." The cell may at this time begin moving about within the matrix, pushing into one filamentous outgrowth, retreating, and then pushing into another until the filaments have been enlarged into "slimeways" broad enough to permit passage of the spindle cells through them. Cell divisions also begin at this time, after which the resultant daughter cells behave in similar fashion. Cytochemical tests show that mucopolysaccharide is associated with the slimeways (Pokorny, 1971).

During these developments, the hyaline filamentous extensions continue to appear and soon become branched and anastomosed to form the ectoplasmic net. The multiplying spindle cells move out radially into the slimeways, eventually gliding along in chains, sometimes several cells abreast. At this time they move mostly toward the periphery of the expanding network (Fig. 239). The advancing front of the ectoplasmic net contains no cells and appears either as anastomosing filaments (Fig. 240) or a continuous band (Fig. 239).

There has been a great deal of discussion about the origin and nature of the ectoplasmic net—whether it develops from filopodial extensions of the cell and contains living matter or whether it is extracellular and nonliving. These questions can best be discussed in Section V, on ultrastructure, but it is useful to note at this point that the uncertainty about whether the cells of *Labyrinthula* move on or within the slimeways has been resolved. Recent ultrastructural studies have shown that the

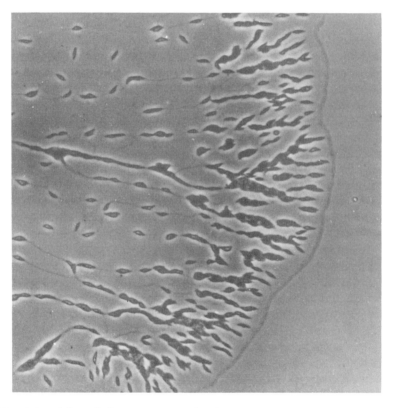

Fig. 239. *Labyrinthula* sp. Margin of colony on agar surface, showing cells passing through channels of ectoplasmic net. Note continuous anterior margin. Phase contrast ×250. From Porter (1969).

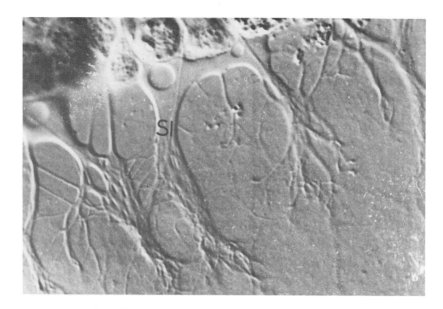

Fig. 240. *Labyrinthula* sp. Edge of colony on agar, showing advance ectoplasmic elements. Nomarski contrast micrograph ×2000. From Pokorny (1971).

cells are contained within the membrane-bounded slimeways, as claimed earlier by Valkanov (1929). However, the mechanism of cell movement has still not been satisfactorily explained. Groups of cells in the slimeway glide along all at the same rate, in some cases as rapidly as 200 μ/minute. The cells become constricted or bent in an impassive manner during movement, only the hyaloplasmic anterior end of the cell showing some pseudopodial activity (F. O. Perkins, personal communication). Single cells removed from their matrix do not move. In thraustochytrids the cells do not glide through the ectoplasmic net, although the net elements are involved in some unknown manner in the slow movement of individual cells over the substrate. In *Labyrinthuloides* such movement occurs over an extended period (Perkins, 1973a). F. O. Perkins and co-workers (personal communication) have recently found that under certain culture conditions, young cells of *Thraustochytrium* and *Schizochytrium* may also show limited movement by means of the ectoplasmic net. However the cells shortly settle down and enlarge into sporangia with rhizoid-like ectoplasmic extensions at their bases (Fig. 241).

In both *Labyrinthula* and the thraustochytrids the ectoplasmic nets function in obtaining nutrients for the protoplasts. Enzymes reach the substrate via their channels, and extracellular products of digestion diffuse back through them to the cells. In thraustochytrids the nets also

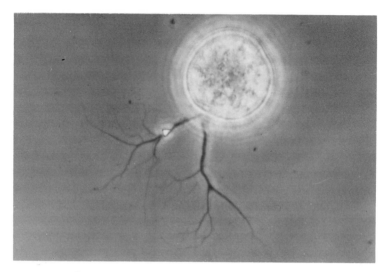

Fig. 241. *Thraustochytrium motivum;* developing sporangium with rhizoidlike ectoplasmic system. Phase contrast ×570. From Perkins (1972).

anchor the developing sporangia to the substrate. In *Labyrinthula* it has been suggested that the slimeways may also serve as a medium for the transmission of stimuli affecting movement patterns in the colony in connection with colony development and cell migration toward a food source (Aschner and Kogan, 1959; Porter, 1969, 1972). Nothing is known about the chemotactic agents that may be involved.

Spindle cell division is mitotic and binary. The nuclear division is intranuclear, with spindle microtubules penetrating through localized breaks in the nuclear envelope at the poles (Porter, 1972). Porter finds that, as the daughter nuclei reorganize, a transverse cell plate develops across the middle of the cell but soon becomes oblique or nearly vertical (as often described in the literature) during elongation of the daughter cells into mature spindle cells.

In at least several species of *Labyrinthula* the vegetative cells accumulate in groups within the ectoplasmic net, become spherical or broadly oval, and secrete thin surrounding walls. In *L. macrocystis* a pellicle is formed around each group of encysted cells and delimits a kind of sorus. When the pellicle breaks down later, small rounded cells, thought to be precursors of spindle cells, are liberated (Young, 1943). The same species reportedly forms individual cysts in which cell division occurs to produce four spherical cells, the fate of which has not been determined (Cienkowsky, 1867). It was of particular interest, therefore, when Hollande and Enjumet (1955) reported that in *L. algeriensis* each cell of the sorus functions as a sporocyte, its contents dividing into several cells that emerge as laterally (or subapically) biflagellate planonts. A similar but unidentified isolate studied by Amon and Perkins (1968) produced sporocytes (which may now be called zoosporangia) that liberated eight laterally biflagellate cells each. Synaptonemal complexes observed at prophase of the first division in the zoosporangium virtually establish the first two of three nuclear divisions in that cell as meiotic. It is of particular interest to note that division of the zoosporangial protoplast is by successive bipartition (protoplast division immediately following each nuclear division) as in certain algal protists, rather than by progressive cleavage, which is characteristic of fungi and myxomycetes.

In the *Labyrinthula* isolate studied by Perkins and Amon (1969) the zoospore has one long anteriorly directed flagellum with mastigonemes and a posteriorly directed smooth one about two-thirds as long. There is an orange stigma or "eye spot," apparently unique to this genus, at the base of the flagellar apparatus. Zoosporulating cultures may have an orange color because of this pigment. Perkins and Amon failed to observe copulation between the presumably haploid cells from the zoosporangium, noting instead that when a zoospore settles onto a substrate it develops directly into a spindle cell. They suggest that copulation may occur between haploid spindle cells. With this in mind, it is interesting to note that Watson (1957) occasionally observed binucleate vegetative cells in *L. vitellina* var. *pacifica* and cells of *Labyrinthuloides minuta* in which the nucleus contained two nucleoli each. He considered these findings evidence of karyogamy.

Although thraustochytrids look quite different from the labyrinthulas, recent investigations, including fine structure studies, indicate that they are related. Both have ectoplasmic nets of similar origin and zoospores with similar flagellation. Thraustochytrids have chytridlike thalli with rhizoidlike ectoplasmic nets (Figs. 241 and 248). A sexual process has not been demonstrated in the life cycle. As in *Labyrinthula*, zoosporogenesis by successive bipartitioning has been described. Further similarities and distinctions are discussed in Section V.

IV. CLASSIFICATION

Whittaker (1969) recognized the phylum Labyrinthulomycota in the kingdom Fungi. Since we do not consider them to be fungi, we have removed them from that kingdom and substituted the name Labyrinthulina for this subphylum of the Gymnomyxa. They may be characterized as protists, primarily of marine or estuarine habitats, which produce ectoplasmic nets by means of special organelles for which we are adopting the term "sagenogens," a shortening of the term "sagenogenetosomes" (net-forming bodies) of Perkins (1972). The cells are essentially non-amoeboid and their nutrition is osmotrophic. Cytokinesis in the sporangium is typically by successive bipartitioning of the protoplast. Zoospores, when present, have an anteriorly directed mastigonemate flagellum and a posteriorly directed smooth one.

Although often associated with the Mycetozoa, the Labyrinthulina are considered unrelated to any group in that subphylum or to the Plasmodiophorina. As treated here, the subphylum contains the single class Labyrinthulea. The signal discovery by Perkins (1972) that the "rhizoidal system" of thraustochytrids is actually an ectoplasmic net produced by sagenogens as in *Labyrinthula* and the similarities between zoospores of the two groups are the prime reasons for our acceptance of his recommendation that labyrinthulas and thraustochytrids be classified together. A comparison of ribosomal RNA molecular weights in the Labyrinthulida by Porter (1974) shows a rather gradual range of values that offers no conflict to the classification of *Labyrinthula* and the thraustochytrids in the same major group and indicates a somewhat closer relationship of *Labyrinthuloides* to the thraustochytrids. The single order Labyrinthulida and two families, the Labyrinthulidae and the Thraustochytriidae, are recognized. The elevation of the latter group to ordinal rank by Sparrow (1973) does not appear justified by recent discoveries. Alterations of our classification may be required as new taxa are discovered and additional information on the life cycles, biochemistry, and ultrastructure becomes available.

Key to the Families and Genera of Labyrinthulida

Trophic stage consisting of spindle-shaped cells that glide
through channels of an ectoplasmic net . . . .   Labyrinthulidae
Single genus: *Labyrinthula*
Cells not typically spindle-shaped and not gliding through ecto-
plasmic net elements . . . .   Thraustochytriidae
    Trophic stage consisting of oval cells that have an extended
    period of migration . . . .   *Labyrinthuloides*
    Trophic stage consisting of cells that enlarge into sporangia
    with basal rhizoid-like ectoplasmic net; cell migration,
    when present, confined to early stages
       Sporangia producing biflagellate planonts
          Thallus developing into a single zoosporangium
             Zoospores liberated through an irregular break in the
             thin sporangial wall or by dissolution of the wall . . .   *Thraustochytrium*
             Zoospores usually liberated through an apical circular
             pore, sporangial wall thicker . . . .   *Japonochytrium*
          Thallus subdividing into zoospore-producing segments .   *Schizochytrium*
       Sporangia producing aplanospores . . . .   *Aplanochytrium*

A. Labyrinthulidae Haeckel (1868)

Some idea of the considerable variety of systems proposed for the classification of this group may be found in the reviews of this subject by Pokorny (1967, 1971), to which may be added the establishment of the new subclass Labyrinthulia in the class Sarcodina by Levine and Corliss (1963). In view of revelations from recent ultrastructural and histochemical studies on this group, detailed discussions of previous taxonomic treatments, of necessity based on less adequate information, seem irrelevant. At present, there appears to be only a single well-defined genus in this family.

1. *Labyrinthula* Cienkowsky (1867)

This is a genus of about seven or eight species.* Both Johnson and Sparrow (1961) and Pokorny (1967) have listed the species and characteristics of all labyrinthulas described before their publications. It is obvious that major emphasis has been placed on cell size in species descriptions. However, variability and degree of overlap in cell size make this an unreliable character. The resulting difficulties in identification have caused many recent investigators to refer to their isolates as *Labyrinthula* spp. A careful monographic study of the genus is needed in which single-cell cultures are used and all helpful morphological and developmental characteristics are employed in making species distinctions. Cultures considered typical of each species should be maintained for distribution to all interested investigators.

A few species of *Labyrinthula* do appear to have quite distinctive characteristics. Both *L. algeriensis* Hollande and Enjumet (1955) and *L. vitellina* Cienkowsky (1867) have biflagellate planonts with orange stigmata that in sporulating cultures of the latter are reported to give the colonies an orange color. However, the vegetative cells of *L. vitellina* also appear to be orange to red in mass, whereas those of *L. algeriensis* are hyaline. Vegetative cells of *L. coenocystis* Schmoller (1960) are pale green individually and dark green in mass. *L. vitellina* var. *pacifica* Watson (1957) and an unidentified isolate discussed by Pokorny (1967) were found to produce, in addition to typical colonies, some irregular, presumably multinucleate plasmodium-like protoplasts of limited size.

* Two newly described species, *Labyrinthula saliens* Quick (1974) and *L. thaisi* Cox and Mackin (1974), produce sporangia with numerous cells and have vegetative cells that move sporadically by means of, but not through, ectoplasmic extensions. They should therefore be reclassified among the thraustochytrids. There is some question as to whether the description of *L. thaisi* is based on a single organism.

Little else is known about the latter, although Watson thought they might occasionally ingest smaller cells.

Differences have also been observed among species with regard to the development of encysted cells; whether large or small, hyaline or pigmented sori are formed; and whether the cysts later give rise to spindle cells directly or undergo nuclear and cell division and produce cells that may or may not become flagellate. Hardly any member of the genus has been studied in sufficient detail to permit a meaningful discussion of how much variability in these characteristics occurs within a single species. The resultant uncertainties about the life cycles and characteristics have led both to the description of new species that are most likely synonymous with earlier described ones and to a hesitancy to describe new isolates that may actually represent new species.

Four other species of *Labyrinthula* have been reported in Europe and America. *Labyrinthula cienkowskii* Zopf (1892), found in strands of *Vaucheria sessilis* in Germany, is probably the only valid freshwater species. *Labyrinthula macrocystis* Cienkowsky (1867) has been reported on Zosteraceae (eel grass family) and a variety of marine algae in Europe and the United States, *L. valkanovii* Karling (1944) on algae in the Black Sea, and *L. roscoffensis* Chadefaud (1956) on marine algae in France.

Labyrinthula zopfii Dangeard (1910), reported to be a freshwater form occurring on *Chlamydomonas* in France, is clearly not a valid species. Dangeard failed to distinguish between an ectoplasmic net with spindle cells and an unstable network of amoeboid cells joined by conventional pseudopodial extensions, as in his species. In addition, the amoebae have the uncharacteristic habit of ingesting portions of the algal protoplasts. Likewise, *L. chattonii* Dangeard (1932) is not an acceptable species, because its description appears to have been based on both a *Labyrinthula* and a proteomyxoid organism (Karling, 1944).

One of the most recently described labyrinthulans is *Labyrinthodictyon magnificum* Valkanov (1969, 1972) from the Black Sea. It was found to grow on marine diatoms in the laboratory. Valkanov considered his new genus distinct from *Labyrinthula* on the basis of its larger masses of spindle-shaped cells and its mode of aggregation leading to the formation of large sori of encysted cells. The spindle cells are capable of gliding centrifugally or centripetally. During aggregation, thick streams of cells gliding through channels of the ectoplasmic net converge in a number of centers to form white stellate masses (called fruiting bodies), in which large numbers of cells become spherical and develop into cysts. The aggregates begin developing near the center of a colony, which may exceed 50 cm in diameter, and new ones appear successively in a peripheral direction. They also develop along the line of juncture between two adjacent colonies. There can be little question that this is a new species, but most of Valkanov's distinctions between his new genus and *Labyrinthula* seem only a matter of degree with regard to certain characteristics instead of new and unique features. The organism needs further study before it can be accepted as representative of a new genus. For the present, the following classification is proposed: *Labyrinthula magnifica* (Valk.) Olive, comb. nov.

2. Doubtful Genera

Several other genera have from time to time been considered members of the Labyrinthulidae. One of these is *Labyrinthomyxa*. The type species, *L. sauvageaui* Dubosq (1921), on the brown alga *Laminaria*, is almost certainly based on the description of two or more organisms, probably including a *Labyrinthula*. Under these circumstances the genus and species cannot be considered valid. A second species, described as *Labyrinthomyxa pohlia* Schmoller (1966, 1971) and said to possess both spindle cells in an ectoplasmic net and limax-type, holozoic amoebae, requires reinvestigation to determine whether two organisms may have been involved in its description. Should it prove to be a valid species, it must be reclassified.

The oyster parasite first described as *Dermocystidium marinum* Mackin, Owen, and Collier was later reclassified as *Labyrinthomyxa marina* (Mackin, Owen, and Collier) Mackin and Ray (1966) on the basis that at certain stages in the life cycle it has small cells that glide along in mucoid tracks. Ultrastructural studies by Perkins (1974b), who failed to find an ectoplasmic net or the organelles that produce it, have led him to exclude the organism from the Labyrinthulida. Its relationship remains uncertain. In addition to the foregoing references, details of the life cycle and ultrastructure of this organism may be found in a number of other publications (Perkins, 1968, 1969; Perkins and Menzel, 1966, 1967).

An unusual organism found on horse dung in Russia and described as *Diplophrys stercorea* Cienkowsky (1876) was thought by its discoverer to be related to *Labyrinthula*. We have collected it on several occasions, mostly on cow dung, in the mountain and piedmont sections of North Carolina. On the natural substrate it produces glistening yellow masses (sorocarps?) that superficially resemble the sorocarps of the cellular slime mold *Guttulinopsis vulgaris* except for color. Each mass consists of numerous spindle-shaped cells in a copious mucilaginous matrix. Each cell has a central uninucleolate nucleus with nearby pigment granule. Cells placed in water soon begin to produce at each end slender filopo-

dium-like processes that resemble the ectoplasmic elements of the Labyrinthulida. Recent studies in our laboratory have shown that the cells
use the ectoplasmic net elements to move in a manner somewhat similar
to that of *Labyrinthuloides* (M. J. Dykstra, unpublished data). We have
cultured the organism for over a year on sterilized cow or horse dung, in
company with one or two bacteria isolated with it that appear necessary
for its maintenance. Studies of the life history and ultrastructure are required to determine whether *D. stercorea* is related to the Labyrinthulida,
and these investigations are now underway in our laboratory.

B. Thraustochytriidae Sparrow (1943, 1960)

Thraustochytrids produce epibiotic sporangia with endobiotic rhizoidlike ectoplasmic extensions that tend to occur in the form of an ectoplasmic net (Figs. 241 and 248). These extensions serve in anchorage and
nutrition of the sporangium and to a limited extent in cell motility. Zoospores are produced either by successive bipartition or progressive cleavage and are laterally biflagellate, the anteriorly directed flagellum being
mastigonemate. Each zoospore is able to generate a new sporangium. One
genus has aplanospores. Sexual reproduction is unknown. Thraustochytrids
are common and widely distributed in marine and estuarine waters, where
they occur as saprobes on plant detritus and possibly as weak parasites
of marine algae. Four genera and a total of 14 species have been described. Gaertner (1972) has reviewed their classification and has proposed a standard method of identification based on growth on pine
pollen in sea water. Both the biology and taxonomy of the group have
been reviewed by Goldstein (1973). We have included *Labyrinthuloides*
in the family along with the four previously recognized genera.

1. *Labyrinthuloides* Perkins (1973a)

This recently described genus has mostly oval or somewhat amoeboid
uninucleate vegetative cells that produce ectoplasmic extensions which
do not enclose the cells. The cells also move about relatively slowly
and only for short distances at a time, frequently reversing themselves.
They move independently rather than in streams or groups. A colony
superficially similar in appearance to that of *Labyrinthula* is characteristically produced in culture. The slow gliding or "spiderlike" motion is
thought to occur by means of changes in position and length of the
ectoplasmic extensions that in some unknown manner enable the cells
to move across the substrate.

Labyrinthuloides yorkensis Perkins (1973a), isolated from estuarine
water and detritus in Virginia, and *L. minuta* (Watson and Raper)
Perkins (1974c), occurring on marine algae in the United States, are

the only known species. In both, the ectoplasmic nets formed by single cells may anastomose where they come into contact (Perkins, 1973b). Typical biflagellate planonts occur in *L. yorkensis*, and multinucleate amoeboid protoplasts that develop from uninucleate cells and later segment into uninucleate vegetative cells have been observed, in both species. Vegetative multiplication in *L. yorkensis* is commonly by means of binary fission or by sporangial formation. In the latter process, a cell becomes spherical and enlarges; nuclear division occurs and the protoplast undergoes successive bipartitioning to form 2–32, sometimes 64, biflagellate zoospores similar to those of *Labyrinthula* in external appearance. It is also claimed that zoospores may occasionally develop by cleavage following the completion of nuclear divisions in the zoosporangium.

In *L. minuta* vegetative cell division occurs in two successive steps. The cell first divides to form two daughter cells; then each daughter cell divides to form a group of four cells surrounded by a thin wall. Multinucleate protoplasts, apparently resulting from mitosis without cytokinesis, are also produced. These may later give rise to uninucleate and plurinucleate protoplasts by plasmotomy (Watson and Raper, 1957). The report by Watson (1957) that posteriorly biflagellate zoospores may be formed by this species requires further investigation.

2. *Thraustochytrium* Sparrow (1936) emend. Johnson (Johnson and Sparrow, 1961)

In the original description, based on the type species *T. proliferum* Sparrow, found on marine algae in coastal waters of the United States and Europe, the sporangium was characterized as opening by an irregular rupture of the wall. As new species were discovered, the mode of zoospore discharge was found to be more versatile. Gaertner (1972) segregated the 11 known species* into three groups: (A) five species in which sporangia proliferate within the empty walls of the old ones (*T. proliferum, T. motivum* Goldstein, *T. aureum* Goldstein, *T. multirudimentale* Goldstein, *T. kinnei* Gaertner); (B) five species not showing sporangial proliferation and with the sporangial wall persistent or deliquescent (*T. roseum* Goldstein, *T. pachydermum* Scholz, *T. striatum* Schneider, *T. globosum* Kobayasi & Ookubo, *T. aggregatum* Ulken); and (C) one species in which the sporangial wall deliquesces before cleavage takes place (*T. visurgensis* Ulken). Following wall deliquescence in the last-mentioned species, the sporangial protoplast changes shape in amoeboid fashion, and when the zoospores are cleaved out

---

* Bohnweg and Sparrow (1974) have described four new species from Antarctic waters: *T. antarcticum, T. rassii, T. kerguelensis,* and *T. amoebium.*

they are already free to swim away. Two species, *T. aureum* and *T. roseum*, have yellowish to orange sporangia.

3. *Japonochytrium* Kobayasi and Ookubo (1953)

The genus is characterized by having relatively thick-walled sporangia and an ectoplasmic net that is enlarged into an apophysis-like swelling at the base of the sporangium. Motile zoospores are liberated through an apical circular pore in the sporangial wall. The only known species is *J. marinum* Kobayasi and Ookubo on marine algae in Japan.

4. *Schizochytrium* Goldstein and Belsky (1964)

Sporangial development in this genus is distinct from that in other genera. As the sporangial thallus enlarges, its protoplast commonly divides by successive bipartition to form smaller thalli, each of which puts out an ectoplasmic net that penetrates the substrate (e.g., pollen grains). The smaller units are produced as tetrads, which may at any stage form zoospores (and are then considered zoosporangia) or undergo further bipartitioning. Zoospore formation by progressive cleavage has also been reported (Perkins, 1974b). The zoospores, or sometimes amoeboid protoplasts, are released through a tear in the sporangial wall. The only known species, *S. aggregatum* Goldstein and Belsky, was isolated from marine water samples in the vicinity of Connecticut, and its development was studied both on pollen grains in seawater and in agar culture.

5. *Aplanochytrium* Bahnweg and Sparrow (1972)

This is the only genus having aplanospores instead of zoospores produced in the sporangia. The mode of sporangial cleavage is uncertain but up to 50 globose aplanospores are formed. These are either liberated by rupture of the sporangial wall or they remain within the intact sporangium and develop *in situ* into new thalli that become clusters of new sporangia. Ectoplasmic nets may begin to develop from aplanospores as early as 20 minutes following liberation from the sporangium. The only known species, *A. kerguelensis* Bahnweg and Sparrow, was isolated from water samples in the vicinity of the Kerguelen Islands in the South Indian Ocean. Its development was studied on pine pollen in sea water cultures.

V. ULTRASTRUCTURE

It has been the ultrastructural approach, more than any other, that has tended to link together *Labyrinthula* and the thraustochytrids, mainly by demonstrating that the slimeways of *Labyrinthula* and the "rhizoids" of thraustochytrids are similar in structure and origin. Most of the fine structure studies on these organisms have been published within the past 6 years.

A. Ultrastructure of *Labyrinthula*

Several species of *Labyrinthula*, some unidentified, have been investigated in considerable detail (Stey, 1968, 1969; Porter, 1969, 1972; Perkins, 1970, 1972, 1973b; Bartsch, 1971; Pokorny, 1971). Hohl (1966) was the first to confirm ultrastructurally the contention of several earlier investigators that the spindle cells are contained within the channels of the ectoplasmic net. Later workers have concurred. Although the spindle cells contain the expected eukaryotic cell organelles, certain of the latter are unusual in their morphology or arrangement within the cell (Fig. 242). One, two, or possibly several large Golgi bodies, each consisting of four to eight flattened stacks of cisternae, have been reported. Porter (1969) states that there are two, each of which delimits small vesicles on both sides. Pokorny (1971), whose admirably detailed study indicates a single large Golgi body somewhat curved around one side of the nucleus and extending three-fourths the cell length, interprets her figures as showing the blebbing of vesicles along the outer face, in addition to the probable fusion of blebs from rough endoplasmic reticulum (ER) and from the apposing face of the nucleus with the inner face of the Golgi body (Fig. 242). Unlike others, who have reported a number of mitochondria, she suggests that there may be only a single large, branched mitochondrion in the cell. The mitochondrion contains rather irregular tubular cristae. The cell also contains numerous lipid globules, rough and smooth ER, and miscellaneous membrane-bound vesicles. Just outside the plasmalemma is found a thin "pellicle" of unknown nature (Pokorny, 1971). It may be homologous with the more conspicuous wall of *Labyrinthuloides* (Perkins, 1972) and other thraustochytrids, in view of the evidence of Perkins (1974c) that it is composed of cell plates (scales).

The truly unique organelle found in these organisms is the sagenogen, without which it is doubtful whether any species should be placed in the Labyrinthulida. The organelle was first named "bothrosome" by Porter (1969), a term denoting the pitlike structure of the organelle in the cell surface of *Labyrinthula*. The later discovery of apparently homologous structures that are not pitlike in thraustochytrids led Perkins to create the term "sagenogenetosome," a shorter and less awkward version of which we have adopted here. Although certain disagreements

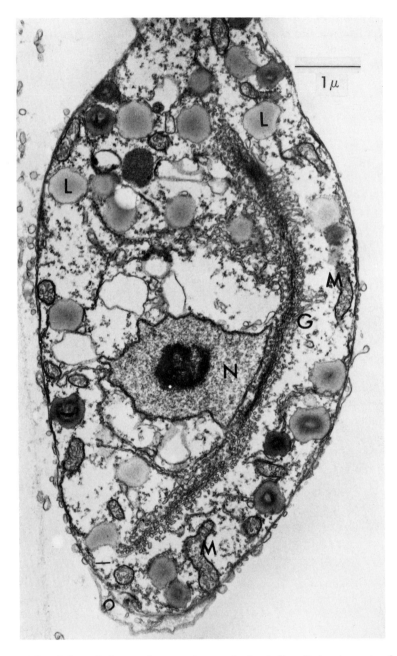

Fig. 242. *Labyrinthula* sp.; electron micrograph of spindle cell showing uninucleolate nucleus (N), large Golgi apparatus (G), lipid bodies (L), and mitochondria (M). From Pokorny (1971).

exist among various authors on the exact relationship of the cells to the ectoplasmic net, it is generally conceded that the latter is a product of the sagenogens and that these remarkable organelles play a part in the movement of labyrinthulan cells. In *Labyrinthula* there are up to 25 sagenogens per cell.

Some idea of the relationship of the vegetative cells to the ectoplasmic net may be obtained by referring to the illustrations (Figs. 243 and 244) from Porter (1972) and Perkins (1974). As noted in Section III, when single spindle cells are placed on a suitable substrate, the sageno-gens soon begin to extrude hyaline sacs (Fig. 238). These sacs are

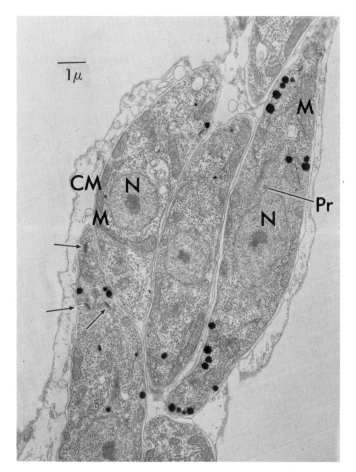

Fig. 243. *Labyrinthula* sp.; electron micrograph of spindle cells within an ecto-plasmic net channel. Note nuclei (N), mitochondria (M), protocentriole (Pr), sagenogens (arrows), and channel matrix (CM). From Porter (1972).

234

7. Labyrinthulina

bounded by a unit membrane. As they enlarge and come into contact, they fuse and thereby enclose the cell in a two-membrane system in which the outer and inner membranes are separated by the hyaline contents of the original sacs. The inner membrane of the sac, which later becomes the inner membrane of the ectoplasmic net, is apparently continuous with the cell plasmalemma in the area of the sagenogens (Fig. 244).

At this point there is some difference of opinion with regard to further

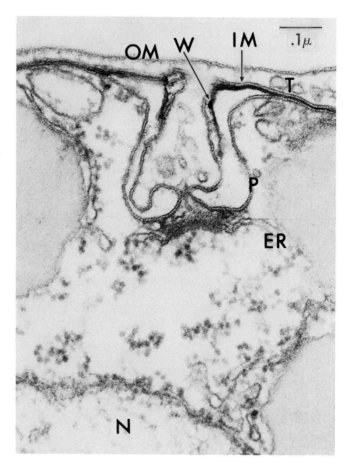

Fig. 244. *Labyrinthula coenocystis;* ultrastructure of the sagenogen, showing continuity between plasmalemma (P) and inner membrane (IM) of ectoplasmic net, outer membrane (OM) of ectoplasmic net, tubular configurations (T) in channel matrix, endoplasmic reticulum (ER), thin cell wall (W), and nucleus (N). From Perkins (1974a); courtesy of Dr. Hiltrud Schwab-Stey.

development of the net. Stey and Pokorny believe that as additional cells become involved in colony development both outer and inner membranes surrounding them fuse, whereas Porter and Perkins conclude that only the outer membrane becomes continuous, each cell in the colony being surrounded individually by the inner membrane. These differences of opinion must be resolved before the question of how the cells move within the ectoplasmic net can be answered.

The sagenogen has a very characteristic structure (Figs. 243, 244). Within the pitlike depression the inner ectoplasmic net membrane appears to be continuous with the plasmalemma. At the base of the pit is an electron-dense plaque surrounded by an electron-dense ring. Profiles of ER are found passing between the basal plaque and the Golgi apparatus. Pokorny concluded that this ER and the Golgi-derived "catenate vesicles" found fusing with the plasmalemma in the area of the pits were transporters of acid mucopolysaccharides to the net elements and that these vesicles were also the source of new membrane for the net. Membrane-bound vesicular elements or cisternae proliferate from the sagenogens into the channels of the ectoplasmic net (Porter, 1969; Perkins, 1973b). Perkins has also shown that the ectoplasmic nets of both *Labyrinthula* and the thraustochytrids proliferate over animal, plant, or fungal cells, often dissolving pores in the cell walls and digesting the cell contents.

It now seems likely that the ectoplasmic net originates as outgrowths from the plasmalemma of the cell. However, the sagenogens are such effective filters that they permit no organelles, not even ribosomes, to pass from the cell into the net elements. Nonetheless, the ectoplasmic system shows certain characteristics of pseudopodia observed in some mycetozoans and other protists. For example, the network sends out at its advance margins filose extensions that may elongate, branch, and anastomose, thereby increasing the area of the net. Under the circumstances, it seems reasonable to consider the ectoplasmic extensions a kind of highly specialized filopodium. Thus far, in conjunction with the sagenogen, they represent a unique type of structure with little else for comparison. Although the nature of cell movement in *Labyrinthula* remains obscure, its eventual explanation will probably be related in large part to the remarkably active contribution of new membranes and mucopolysaccharides to the ectoplasmic net via the sagenogens.

The ultrastructure of mitotic cell division has been studied by several investigators (Stey, 1969; Perkins, 1970; Porter, 1972). There is no evidence of centrioles during interphase but during mitosis two electron-dense aggregates with a nine-spoke cartwheel structure—the "protocentrioles"—appear in slight depressions on opposite sides of the nucleus.

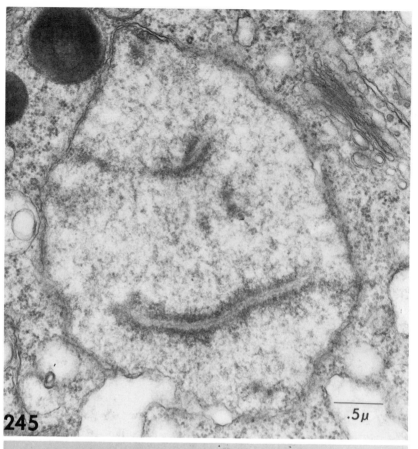

245   .5µ

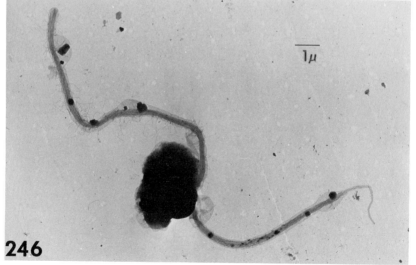

246

Microtubules emanating from them penetrate into the nucleus through polar breaks in the nuclear envelope. Following telophase the nuclear envelope is regenerated in the same areas. Cytokinesis occurs by means of a plate of anastomosing vesicles across the cell, and as the apposing ends of the two daughter cells elongate, the plane of division shifts from transverse to the more frequently observed diagonal position. Sagenogens then develop *de novo* along the apposing diagonal cell membranes and produce ectoplasmic extrusions that eventually extend across the ends of the two daughter cells and fuse with the preexisting inner ectoplasmic membrane around the cells, after which the cells are ready to separate (Porter, 1972).

The first convincing evidence of meiosis in the life cycle of *Labyrinthula* was obtained by Perkins and Amon (1969) and Perkins (1970) in an unidentified isolate. Zoosporulation in the rounded cells was preceded by the appearance of synaptonemal complexes (Fig. 245) in the prophase I nucleus and the *de novo* development of a pair of centrioles, each from a pair of procentrioles, at the incipient spindle poles. A study of serial sections through a meiotic prophase nucleus demonstrated the presence of nine distinct synaptonemal complexes ($n = 9$), both ends of which were attached to the nuclear envelope (Moens and Perkins, 1969). Spindle microtubules connected the centrioles outside the nucleus before fragmentation of the nuclear envelope. The second meiotic division was followed by a third, equational division. The sporangial protoplast underwent successive bipartition, dividing after each nuclear division, and the eight resultant uninucleate products developed into laterally biflagellate, stigmate zoospores. The sporangial wall was no more than a thin pellicle similar to that around the vegetative cell. The kinetosomes were derived directly from the centrioles. The mastigonemes of the anteriorly oriented flagellum each had two very fine extensions at their tips. The zoospore of *L. vitellina* in shadowed whole mount is shown in Fig. 246.

B. Ultrastructure of Thraustochytrids

Most of the fine structure studies of thraustochytrids have been concentrated on the sagenogens (Fig. 247), ectoplasmic nets (Fig. 248), and

Fig. 245. *Labyrinthula* sp.; electron micrograph of nucleus, showing parts of two synaptonemal complexes. From Perkins and Amon (1969).

Fig. 246. *Labyrinthula vitellina;* shadowed whole mount of zoospore, showing mastigonemate anterior flagellum and posterior whiplash flagellum. From Porter (1974).

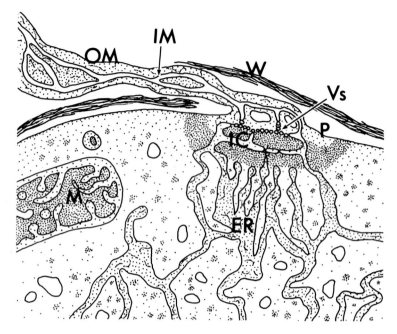

Fig. 247. *Labyrinthuloides minutum;* diagram of sagenogen, showing endoplasmic reticulum (ER) continuous with inner cisternae (IC) of sagenogen, layer of small vesicles (Vs), internal membrane (IM) and outer membrane (OM) of net element, mitochondrion (M), plasmalemma (P), and cell wall (W) of overlapping scales. From Perkins (1972).

zoospores. The significant discoveries of Perkins (1972) and Harrison and Jones (1974) that members of this family, formerly classified among the Oomycetes, have, instead of true rhizoids, ectoplasmic nets produced by sagenogens, and the discoveries of Darley and Fuller (1970), Darley *et al.* (1973), and Perkins (1974a) on cell wall structure go a long way toward justifying the alliance of thraustochytrids and *Labyrinthula.*

Although the sagenogens of thraustochytrids are not in distinct pits (Fig. 247), their basic structure and relationship to ER in the cell appear essentially the same as in *Labyrinthula,* if somewhat less complex (Perkins, 1972). Cells of *Labyrinthuloides minutum* have two to five sagenogens each and those of *L. yorkensis* have four to nine. Sagenogens occur singly in *Japonochytrium* sp., singly or in a closely related pair in *Schizochytrium aggregatum,* and in clusters of 15 or more in *Thraustochytrium motivum.* The ectoplasmic net of the latter three species typically originates from a single trunk element produced by one or two sagenogens or cooperatively by a cluster of sagenogens. Membrane-bounded tubular

cisternae are produced in the matrix of the ectoplasmic elements, and in
T. motivum (Fig. 249) these cisternae may occur in conspicuous anastomosing aggregates with latticelike structure in sectional view (Perkins,
1972, 1973b).

In a study of *T. aureum*, Goldstein *et al.* (1964) found a Golgi apparatus and a pair of centrioles in the vicinity of the interphase nucleus.
Porter (1974) states that vegetative nuclear division in thraustochytrids
differs from that in *Labyrinthula* by having true centrioles rather than
protocentrioles at the spindle poles, true centrioles in *Labyrinthula* being
present only during meiosis. There has been no evidence, ultrastructural
or otherwise, of meiosis in a thraustochytrid.

The laminated nature of the cell wall in thraustochytrids, first discovered by Goldstein *et al.* (1964), has recently been confirmed in both
Schizochytrium and *Thraustochytrium* and was shown to result from
the presence of overlapping circular scales (Fig. 250) that lack any detectable substructure (Darley *et al.*, 1973). These scales are preformed in
cisternae that are probably of Golgi origin, and they are expelled at
the periphery of the cell where, combined with matrix material, they
contribute to cell wall formation. A similar wall structure has also been

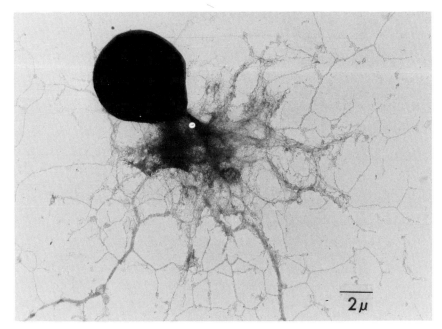

Fig. 248. *Thraustochytrium motivum;* whole mount of a developing sporangium
with rhizoidlike ectoplasmic system at its base. From Perkins (1973b).

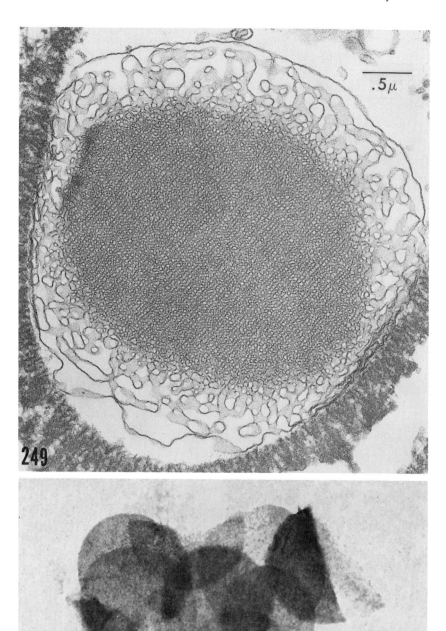

249

250

reported in *Labyrinthuloides* (Perkins, 1974a). There is a single Golgi apparatus in the *Schizochytrium* cell, several in *Thraustochytrium*.

An isolate of *Japonochytrium* resembling *J. marinum* was found to have multilamellate sporangial walls but scales were not detected (Harrison and Jones, 1974). Results obtained with other thraustochytrids suggest that the walls of *Japonochytrium* may also prove to be composed of scales. The zoospores contain a prominent Golgi body and in most respects are ultrastructurally like those of other thraustochytrids.

The zoospores of both *Thraustochytrium* (Kazama, 1974) and *Schizochytrium* (Perkins, 1974a) are also enclosed by a thin layer of similar scales. The zoospores are in most ways similar to those of *Labyrinthula* both in flagellation (Gaertner, 1964) and in having two fine extensions at the tip of each mastigoneme of the anteriorly directed flagellum (Perkins, 1974a). Also, within each kinetosome are found several dense granules or a dense rod as in *Labyrinthula*. Zoospores of thraustochytrids do not appear to have eye spots but those of *Thraustochytrium* contain sagenogens. A study of the ultrastructural cytochemistry of acid phosphatase distribution in the zoospore of *Thraustochytrium* (Kazama, 1973) shows that this enzyme is localized in Golgi saccules and vesicles, multivesicular bodies, ER, and autophagic vacuoles. A large cytolysome develops within the mature zoospore. Additional details on zoospore structure may be found in the publications by Kazama (1972a,b) and Perkins (1974a).

Herpes-like virus particles (nucleocapsids) have been reported by Kazama and Schornstein (1972) in *Thraustochytrium* sp. and by Perkins (1974a) in *Schizochytrium aggregatum*. Infections of the latter often result in cell lysis.

Further research is required to determine whether the recently described genus *Althornia* belongs in the Thraustochytriidae. The single known species, *A. crouchii* Jones and Alderman (1972), has zoospores with peripheral scales, a large Golgi body, and flagellation as in *Thraustochytrium*, but the sporangia, which are also surrounded by scales, apparently lack sagenogens and ectoplasmic nets.

As interest in marine biology continues to expand, a number of new taxa are likely to be added to the Labyrinthulina. These discoveries, in addition to further investigations of life cycles and ultrastructure, should produce a better understanding of the nature and relationships of this unusual group of protists.

Figs. 249 and 250. Ultrastructure of *Thraustochytrium*. Fig. 249. Periodic arrangement of internal membranes in the ectoplasmic net element of *T. motivum*. From Perkins (1973b). Fig. 250. Wall scales of *Thraustochytrium* sp. From Darley *et al.* (1973).

Phylogeny

CHAPTER 8

Phylogenetic Implications

While phylogeny is one of the most controversial of all subjects, it is also one of the most important, since some understanding of it is basic to all natural systems of classification. In such groups as the mycetozoans, which have left little or no fossil record, it is a subject that can be discussed without much fear of contradiction. At the same time, this lack of solid information permits a wide variety of opinions.

In the following discussion we will combine information presented in the foregoing chapters with ideas, old and new, offered by previous investigators to arrive at a tentative concept of phylogeny for the mycetozoans and their associates (Fig. 251). We realize that frequent modification will be required as new information accumulates. It is hoped that the end result will be a better understanding of interrelationships among these organisms and their affinities with outside groups, which will in turn lead to a more natural system of classification than we have heretofore had. If this concept is accepted as a challenge by others to test

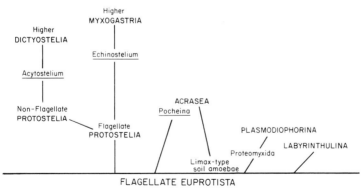

Fig. 251. Phylogeny of the Gymnomyxa: tentative proposal. As indicated, the Acrasea are probably not monophyletic.

its ideas by presenting new evidence, for or against, it will have served its major purpose.

Figure 251 diagrams what we believe to be the most logical concept at this time for the evolution of mycetozoans and associated groups from aquatic flagellate protists, most of which may have been simple algal forms. The Protostelia are the simplest of the eumycetozoans, and their flagellate members are considered the most primitive. The presence of scale-forming vesicles in the flagellate protostelid *Ceratiomyxella* similar to those found in certain groups of flagellate algal protists (Manton, 1967; Brown *et al.*, 1970) suggests the possible origin of protostelids, and therefore of eumycetozoans, from some such group.

The origin of myxomycetes from flagellate protostelids is indicated by the presence of very similar flagellate cells and amoeboid cells with filose pseudopodia, as well as fruiting bodies in the simplest myxomycetes that intergrade in structure with those of protostelids. In fact, it is somewhat difficult to draw a sharp line between a simple myxomycete such as *Echinostelium lunatum*, which commonly has 4- to 8-spored fruiting bodies, and the protostelid genus *Protosporangium*, in which 4-spored fruiting bodies are common. Also, ultrastructural studies have revealed the presence of similar, unique wall projections on the spores of the protostelid *Cavostelium bisporum* and two species of *Echinostelium*, including the minute *E. lunatum*. In *C. bisporum*, as in the myxomycetes, a single mitosis and cytokinesis occur in the sporogenous protoplast just before spore formation. In addition, the flagellate cells of both are remarkably similar under light and electron microscopes. In view of these resemblances, it is thought that *Cavostelium* offers the closest known link between the protostelids and *Echinostelium* of the myxomycetes.

In all five orders of the Myxogastria the spores are formed by cleavage of the sporocarp protoplast, following which meiosis occurs, uniquely, within the maturing spores. Further similarities are found in the general occurrence of precleavage mitosis, spore ornamentation, planonts with uniformly similar flagellar apparatus, and alternation of a uninucleate, unicellular haplophase (myxamoebae and flagellate cells) with a more prominent multinucleate, coenocytic diplophase (plasmodium). In spite of great differences in the size and form of fruiting bodies among myxomycetes, their many basic similarities indicate a monophyletic origin of the group.

If the true slime molds are monophyletic in origin and *Echinostelium* is accepted as the most primitive genus, it follows that higher myxomycetes evolved from it. This means that more complex stalked sporangiate forms developed later, probably to be followed by sessile sporangiate ones, and then by plasmodiocarpous and aethalioid forms from the latter. Support for this viewpoint is found in the pseudoaethalium, a type of sporocarp transitional between a cluster of individual sporangia and a true aethalium. The pseudoaethalium of *Dictydiaethalium,* for example, is comprised of a palisade of sessile, closely appressed, and partially fused sporangia. Intergradations between sporangiate and plasmodiocarpous sporocarps are also known. For example, in certain normally sporangiate species of *Trichia* and *Hemitrichia* it is not uncommon to find small plasmodiocarpous fruiting bodies that have apparently developed through failure of portions of the fruiting plasmodium to complete the delimitation of individual sporangia.

It is quite likely, as Ross (1973) has suggested, that the Stemonitida, a number of which are characterized by endogenous stalk development and other possibly unique features, branched off early from the main line of myxomycete phylogeny. However, a much more intensive study of a greater variety of genera in this and other orders is needed before the relative values of these developmental features can be properly assessed. In *Stemonitis,* as in *Echinostelium lunatum* and the protostelids, the stalk of the sporocarp has maintained its slender, tubular character, whereas in most stipitate myxomycetes it has become thicker and is often filled with encysted protoplasts or with detritus discharged from the plasmodium. In *Stemonitis,* as in *Echinostelium,* protostelids, and dictyostelids, the fruiting bodies are enclosed by a thin membranous sheath. That portion of the sheath enclosing the spores is thought to be the precursor of the peridium of most myxomycetes, which probably evolved with the tendency for additional materials (such as calcium carbonate) to be deposited in or on it (Olive and Stoianovitch, 1971c). Likewise, the discoid base of the sheath in protostelids and *Echinostelium* is probably the precursor of the hypothallus of myxomycetes.

Although the slime molds are often included in the fungi (Martin, 1960; Martin and Alexopoulos, 1969), their phagotrophic mode of nutrition, type of flagellation, amoeboid movement, localization of meiosis in developing spores, and important biochemical differences make it highly unlikely that they are related to any members of that group. Recently, Poulos et al. (1971) reported that plasmodia of *Physarum polycephalum* contain an abundance of ether-linked phospholipids similar to those found in certain euprotists but not in the fungi. They take this as evidence that the myxomycetes are more closely related to protozoans than to fungi. Attempts to homologize the myxomycete slime sheath or the walls of spores and spherules with cell walls of fungi have been of a superficial nature. The sheath—a galactose polymer—and the walls of spores and spherules—comprised mainly of a galactosamine polymer and glycoprotein—have compositions unlike those of cell walls of any known fungi. The evidence indicates that the Myxogastria are a terminal group in the evolution of mycetozoans and that no higher groups have arisen from them.

The dictyostelid cellular slime molds are believed to have evolved from nonflagellate protostelids. *Acytostelium*, characterized by a narrow noncellular stalk tube similar to that of protostelids, is thought to occupy an intermediate position between the two subclasses. This thesis is supported by the discovery of Dutta and Mandel (1972) that the guanine–cytosine (GC) content of *Protostelium irregularis* (35%) is close to that of *Acytostelium leptosomum* (37%) but not to that of the four *Dictyostelium* species examined (22–25%).

It is postulated that evolution of Dictyostelia from the Protostelia was accompanied by the initiation of a chemical attractant system that induces amoebae to come together in aggregates and produce multicellular fruiting bodies. Both Raper (1960) and Olive and Stoianovitch (1969) have noted the tendency of trophic amoebae to clump in species of *Protostelium*, although these cell clusters do not form multicellular fruiting bodies. As in protostelids and myxomycetes, dictyostelid amoebae produce filose pseudopodia but flagella are absent. However, the recent discovery of centriole-like structures that may represent vestigial centrioles or kinetosomes in the amoebae of *Polysphondylium* (U.-P. Roos, personal communication and Fig. 106) suggests that dictyostelids may not be far removed from their flagellate ancestors.

The most primitive dictyostelid appears to be *Acytostelium*, a genus characterized by a narrow, simple, acellular stalk tube bearing an apical sorus of spores. When amoebae ready to commence the fruiting process are plated out on washed agar devoid of nutrients, numerous small fruiting bodies are formed. Some that contain only one or two spores are indistinguishable from the sporocarps characteristic of protostelids. *Dictyostelium* and *Polysphondylium*, with their broader double-layered stalk tubes filled with a framework of dead cells, show a higher degree of differentiation and are thought to have evolved from *Acytostelium*, with the poorly known *Coenonia* possibly branching off the same line. The dictyostelids, like the myxomycetes, appear to have had a monophyletic origin and to have given rise to no higher forms.

Earlier investigators and zoologists in general have long been aware of two major types of amoebae among the protists, i.e., those with filose pseudopodia and those with lobose pseudopodia ("limax" type). The former type occurs throughout the eumycetozoans, whereas limax amoebae are characteristic of acrasid cellular slime molds. E. W. Olive (1902) made this a primary distinguishing feature between his two main families of cellular slime molds (although insufficient information available on *Acrasis* led him to place it among the dictyostelids). Recent investigations have further emphasized differences between dictyostelid and acrasid cellular slime molds. In all acrasids that have been studied in any detail, aggregation occurs not by streaming but by migration, singly or in groups, of amoebae that are not elongated in the direction of movement as they are in dictyostelids. In further contrast to the latter group, no stalk tube is present in the acrasid fruiting body and the stalk cells remain viable. These distinctions can hardly be considered casual ones, because they reflect basic differences in modes of sporogenesis in the two groups and are very likely of phylogenetic significance.

Although the eumycetozoans appear to be interrelated within a single phylogenetic series, it is becoming increasingly doubtful that the same can be said about the Acrasea. For example, there are striking differences between *Acrasis* and *Copromyxa*, especially with respect to sorogenesis. In *A. rosea* the stalk cells are deposited by the globular sorogen as it migrates upward on the elongating stalk, and mature stalk cells and spores are morphologically distinct. However, in *Copromyxa* the sorocarp is reported to grow by the continuous migration of amoebae individually up the sides of the stalk and branches until they reach the sorocarp terminals, where they encyst and thus extend the sorocarps. There are no morphological differences among encysted cells in the sorocarp. The limited information available on sorocarp development in *Guttulinopsis* indicates that it has a pattern distinct from that of either *Acrasis* or *Copromyxa*.

Ultrastructural studies, although still lacking or fragmentary for some groups, have supplemented these findings. Throughout the Eumycetozoa the mitochondria have tubular cristae, whereas the Acrasea are heterogeneous with respect to this feature. The mitochondria of *Pocheina rosea, Acrasis rosea,* and *Guttulinopsis vulgaris* have well-defined disc-shaped cristae, whereas those of *Copromyxa arborescens* have tubular ones. Little effort has so far been made to assess the phylogenetic significance of mitochondrial types in lower organisms. In the fungi the two types appear to be aligned with other important characteristics that sharply distinguish the Oomycetes and Hyphochytridiomycetes from the Eumycota. The former are generally reported to have tubular mitochondrial cristae and the latter, disc-shaped ones, but further studies are needed to determine whether such a distinction is consistent. In any event, future systems of classification of protistan groups should probably take mitochondrial characteristics into account.

Prespore vacuoles, which are characteristic of dictyostelids, have not yet been identified in any of the acrasids, but the search for them in the latter group has been far from exhaustive. In encysted cells of the sorocarp of *Acrasis rosea* the nucleus—and especially the nucleolus—is ultrastructurally more complex than that of any other mycetozoan that has been investigated. The nucleolus is of a complex nature and contains distinctive "lamellate elements" comprised mainly of ribosomes or ribosome precursors. It will be of particular interest to determine whether *Pocheina rosea*, thought to be related to *A. rosea*, has a similar nucleus.

Such forms as *Copromyxa arborescens*, in which microcysts and all cells of the sorocarp are morphologically similar (even at the ultrastructural level), appear to be the simplest of the acrasids, and the genus is probably a primitive one. However, the presence of flagellate cells in *Pocheina rosea* suggests that it too is a primitive genus, in spite of a more highly developed sorocarp with morphologically distinct spores and stalk cells that appear to ally it closely with *Acrasis rosea*. Under the circumstances it is difficult to understand how *Copromyxa* could have evolved from the *Pocheina–Acrasis* lineage. Accordingly, we believe that future studies will favor the polyphyletic origin of acrasids from simpler protists. In the evolution of simple forms, such as *Copromyxa*, it is quite conceivable that the immediate progenitor was a soil amoeba. It should be recalled (Chapter 1) that the ability to form fruiting bodies from cell aggregates has clearly evolved independently on more than one occasion (e.g., myxobacteria, a ciliate, dictyostelids, acrasids), and there is no reason to assume that it may not have occurred more than once in the evolution of acrasid cellular slime molds.

With regard to the Plasmodiophorina and Labyrinthulina, which have often been associated with the mycetozoans, there is a lack of evidence that either is closely related to any group of the Mycetozoa; nor do they appear related to the Fungi, where they have sometimes been placed. As noted in Chapter 6, there are some reasons to suspect that the plasmodiophorids are related to and possibly have evolved from certain of the Gymnococcidae in the order Proteomyxa, which are generally classified with the Euprotista. The most likely progenitors in this family have endobiotic, holozoic, plasmodium-like protoplasts occurring within the cells of algae, as well as zoosporangia and zoospores that resemble those of plasmodiophorids. However, much more information on the cytology and development of these organisms is needed before closer comparisons can be made.

Since their discovery, the labyrinthulas have stimulated discussion with regard to their possible position among other groups of organisms. Similar discussions have involved the later discovered thraustochytrids, which until recently were thought to be related to Oomycetes. In the last several years, ultrastructural and histochemical investigations have tended to ally thraustochytrids with *Labyrinthula* and to show that they are probably unrelated to either the Fungi or Mycetozoa.

Perkins justified the closer association of the two groups primarily on the basis that both possess sagenogens and ectoplasmic nets, as well as similar laterally biflagellate zoospores. On the other hand, Porter (1974) has noted certain distinctions between the two groups, including differences in ribosomal ribonucleic acid (rRNA) molecular weights, which he thinks mitigate against lumping them together in a single family. At the same time, his analysis shows that the rRNA molecular weights of *Labyrinthuloides* are intermediate between those of *Labyrinthula* and *Thraustochytrium*, actually being closer to the latter. Most of the evidence that has accumulated in recent years suggests that labyrinthulas and thraustochytrids are related organisms and should be classified together.

There can now be little question about the propriety of removing the thraustochytrids from the Oomycetes. This is justified by the following discoveries in the thraustochytrids: (1) galactose rather than glucose is the chief monosaccharide component of the cell wall (Darley and Fuller, 1970); (2) the wall is constructed of thin flat scales and matrix material rather than from amorphous deposits; (3) sagenogens and ectoplasmic net rather than a rhizoidal system are present; and (4) the rRNA molecular weights are significantly different from those of Oomycetes. As previously noted, wall scales comparable to those reported in *Thraustochytrium. Schizochytrium*, and *Labyrinthuloides* are now believed to be present in *Labyrinthula*. In addition, successive bipartition of sporangial protoplasts, reported in both thraustochytrids and *Labyrinthula* is not known in Oomycetes, where progressive cleavage is characteristic.

It is equally clear that the Labyrinthulina are unrelated to other groups of organisms described in the foregoing chapters. Most of the differences are already apparent. In addition, the rRNA molecular weights of the Labyrinthulina and Mycetozoa are now known to be significantly different (Porter, 1974). Comparisons between the mycetozoan plasmodium and an ectoplasmic net are obviously of a superficial nature.

The question of the phylogenetic origin of the Labyrinthulina remains unanswered, although some interesting implications have surfaced. Scale formation in these organisms is similar to that reported in certain algal protists (e.g., Manton, 1967; Brown, 1969), in which the scales are formed in Golgi-derived vesicles that transport them to the cell surface. Also, their zoospores are remarkably like those of certain chrysophycean and xanthophycean algal protists, both in structure and mode of swimming. As Perkins (1974b) has noted, a study of the chrysophycean protist,

Phaeaster pascheri Scherffel, by Belcher and Swale (1971) has demonstrated the presence of cells with a long anterior mastigonemate flagellum, a rudimentary flagellum overlying a red stigma, thin peripheral scales, and fine rhizopodia. The possible relationship of the latter to the ectoplasmic net of the Labyrinthulina is suggested by the fact that they sometimes contain tubular profiles and an electron-dense "septum" (precursor of sagenogen?) at the base. Although it is still too early to name a likely progenitor of the Labyrinthulina, further studies of such forms as *Phaeaster pascheri* may implicate some such algal protist in this role.

In summary, it has been tentatively proposed that the phylum Gymnomyxa of the kingdom Protista is comprised of at least four independent phylogenetic series, represented by classes Eumycetozoa and Acrasea of subphylum Mycetozoa and by subphyla Plasmodiophorina and Labyrinthulina. Much more research on these organisms is required in order to test this hypothesis more fully and to obtain a better concept of their progenitors among the Euprotista.

Appendix

The following agar media are recommended for the culture and maintenance of mycetozoans. The pH may need to be adjusted for the particular species being cultured.

1. Cornmeal Agar, Supplemented
   Difco cornmeal agar .... 17 gm
   Dextrose .... 2 gm
   Yeast extract .... 1 gm
   Distilled water .... 1 liter

2. Hay Infusion Agar
   Dry hay .... 2.5 gm
   Agar .... 17.0 gm
   Distilled water .... 1.0 liter

Add the distilled water to the hay and autoclave for 10 minutes. Filter through cheesecloth to remove the hay and return the volume to 1 liter. Add agar, adjust to desired pH with $K_2HPO_4 \cdot 3H_2O$, and autoclave at 15 pounds pressure for 15 minutes.

3. Knop's Solution Agar
   $Ca(NO_3)_2 \cdot H_2O$ .... 0.8 gm
   KNO_3 .... 0.2 gm
   KH_2PO_4 .... 0.2 gm
   $MgSO_4 \cdot 7H_2O$. .... 0.2 gm
   $FeSO_4$ or $FePO_4$ .... trace
   Agar .... 20.0 gm
   Distilled water .... 1.0 liter

4. Lactose–Yeast Extract Agar
   Lactose .... 1.0 gm
   Yeast extract .... 0.5 gm
   Agar .... 15.0 gm
   Distilled water .... 1.0 liter

5. Oak Bark Agar
    Bark of white oak ....   2.5 gm
    Agar ....   7.5 gm
    Distilled water ....   1.0 liter

   Use the outer bark of *Quercus alba* (or another member of the white oak group).
Add water to the bark and autoclave for 10 minutes. Filter off the bark and bring
the volume back to 1 liter. Add agar, adjust to desired pH, and autoclave at 15 pounds
pressure for 15 minutes.

6. Oat-Flake Agar (Methods A and B)
   A. Cover the bottom of a Petri dish with a layer of rolled oats (ca. 5 mm
thick) and cover with 2% melted agar. Autoclave for 30 minutes at 15 pounds pressure.
   B. Autoclave 100 gm of rolled oats in 1 liter of distilled water for 30 minutes.
Filter through cheesecloth, bring filtrate back to 1 liter, add 15 gm of agar, and
autoclave at 15 pounds pressure for 15–20 minutes. Sterile oat flakes may be added
to the agar surface at the time of inoculation with the mycetozoan.

7. Malt-Yeast-Extract Agar
    Malt extract ....   0.1 gm
    Yeast extract ....   0.1 gm
    Agar ....   15.0 gm
    Distilled water ....   1.0 liter

Bibliography

Ainsworth, G. C. (1973). Introduction and keys to higher taxa. *In* "The Fungi" (G. C. Ainsworth, F. K. Sparrow, and A. S. Sussman, eds.), Vol. 4B, pp. 635–648. Academic Press, New York.

Aist, J. R., and Williams, P. H. (1971). The cytology and kinetics of cabbage root hair penetration by *Plasmodiophora brassicae*. *Can. J. Bot.* **49,** 2023–2034.

Aldrich, H. C. (1966). A study of the ultrastructural details of morphogenesis in the myxomycete *Physarum flavicomum*. Ph.D. Thesis, University of Texas, Austin.

Aldrich, H. C. (1967). The ultrastucture of meiosis in three species of *Physarum*. *Mycologia* **59,** 127–148.

Aldrich, H. C. (1968). The development of flagella in swarm cells of the myxomycete *Physarum flavicomum*. *J. Gen. Microbiol.* **50,** 217–222.

Aldrich, H. C. (1969). The ultrastructure of mitosis in myxamoebae and plasmodia of *Physarum flavicomum*. *Amer. J. Bot.* **56,** 290–299.

Aldrich, H. C. (1970). Pre- and postmeiotic events in spores of the myxomycete *Didymium iridis*. *J. Cell Biol.* **47,** 4a (abstr.).

Aldrich, H. C. (1974). Spore cleavage and development of wall ornamentation in two myxomycetes. *Proc. Iowa Acad. Sci.* **81,** 19–26.

Aldrich, H. C., and Gregg, J. H. (1973). Unit membrane structural changes following cell association in *Dictyostelium*. *Exp. Cell Res.* **81,** 407–412.

Aldrich, H. C., and Mims, C. W. (1970). Synaptonemal complexes and meiosis in myxomycetes. *Amer. J. Bot.* **57,** 935–941.

Alexopoulos, C. J. (1960a). Gross morphology of the plasmodium and its possible significance in the relationships among the myxomycetes. *Mycologia* **52,** 1–20.

Alexopoulos, C. J. (1960b). Morphology and laboratory cultivation of *Echinostelium minutum*. *Amer. J. Bot.* **47,** 37–43.

Alexopoulos, C. J. (1962). "Introductory Mycology." Wiley, New York.

Alexopoulos, C. J. (1963). The myxomycetes II. *Bot. Rev.* **29,** 1–78.

Alexopoulos, C. J. (1964). The rapid sporulation of some myxomycetes in moist chamber culture. *Southwest. Natural.* **9,** 155–159.

Alexopoulos, C. J. (1966). Morphogenesis in the myxomycetes. *In* "The Fungi" (G. C. Ainsworth and A. S. Sussman, eds.), Vol. 2, pp. 211–234. Academic Press, New York.

Alexopoulos, C. J. (1969). The experimental approach to the taxonomy of the myxomycetes. *Mycologia* **61,** 219–239.

Alexopoulos, C. J. (1973). The myxomycetes. *In* "The Fungi" (G. C. Ainsworth, F. K. Sparrow, and A. S. Sussman, eds.), Vol. 4B, pp. 39–60. Academic Press, New York.

Alexopoulos, C. J., and Brooks, T. E. (1971). Taxonomic studies in the myxomycetes. III. Clastodermataceae: A new family of the Echinosteliales. *Mycologia* **63**, 925–928.

Alexopoulos, C. J., and Zabka, G. G. (1962). Production of hybrids between physiological races of the true slime mould *Didymium iridis. Nature (London)* **193**, 598–599.

Alléra, A., and Wohlfarth-Bottermann, K-E. (1972). Weitreichende fibrillare Protoplasmadifferenzierungen und ihre Bedeutung für die Protoplasmaströmung. IX. Aggregationszustände des Myosins und Bedingungen zur Entstehung von Myosinfilamenten in den Plasmodien von *Physarum polycephalum. Cytobiologie* **6**, 261–286.

Amon, J. P., and Perkins, F. O. (1968). Structure of *Labyrinthula* sp. zoospores. *J. Protozool.* **15**, 543–546.

Anderson, J., Fennel, D., and Raper, K. B. (1968). *Dictyostelium deminutivum,* a new cellular slime mold. *Mycologia* **60**, 49–64.

Andresen, N., and Pollock, B. M. (1952). A comparison between the cytoplasmic components in the myxomycete, *Physarum polycephalum,* and in the amoeba, *Chaos chaos. C. R. Trav. Lab. Carlsberg, Ser. Chim.* **28**, 247–264.

Archer, W. (1875). On *Chlamydomyxa labyrinthuloides. Quart. J. Microsc. Sci.* **15**, 107–130.

Arnaud, G. (1948). Les Heimerliacées, subdivision des Acrasiales? *C. R. Acad. Sci.* **226**, 1744–1746.

Arnaud, G. (1949). Les Heimerliacées, subdivision des Acrasiales? *Botaniste* **34**, 35–55.

Aschner, M., and Kogan, S. (1959). Observations on the growth of *Labyrinthula macrocystis. Bull. Res. Counc. Isr., Sect. D* **8**, 15–24.

Ashworth, J. M. (1971). Cell development in the cellular slime mould *Dictyostelium discoideum. Symp. Soc. Exp. Biol.* **25**, 27–49.

Ashworth, J. M., and Sackin, M. J. (1969). Role of aneuploid cells in cell differentiation in the cellular slime mould *Dictyostelium discoideum. Nature (London)* **224**, 817–818.

Ashworth, J. M., Duncan, D., and Rowe, A. J. (1969). Changes in fine structure during cell differentiaton of the cellular slime mould *Dictyostelium discoideum. Exp. Cell Res.* **58**, 73–78.

Bahnweg, G., and Sparrow, F. K. (1972). *Aplanochytrium kerguelensis* gen. nov. spec. nov., a new phycomycete from subarctic marine waters. *Arch. Mikrobiol.* **81**, 45–49.

Bahnweg, G., and Sparrow, F. K., Jr. (1974). Four new species of *Thraustochytrium* from Antarctic regions, with notes on the distribution of zoosporic fungi in the Antarctic marine ecosystems. *Amer. J. Bot.* **61**, 754–766.

Bartsch, G. (1971). Cytologische Beobachtungen an *Labyrinthula coenocystis* Schmoller bei verschiedenen Kulturbedingungen. *Z. Allg. Mikrobiol.* **11**, 79–90.

Belcher, J. H., and Swale, E. M. F. (1971). The microanatomy of *Phaeaster pascheri* Scherffel (Chrysophyceae). *Brit. Phycol. J.* **6**, 157–169.

Berry, J. A., and Franke, R. G. (1973). Taxonomic significance of intraspecific isozyme patterns of the slime mold *Fuligo septica* produced by disc electrophoresis. *Amer. J. Bot.* **60**, 976–986.

Bessey, E. A. (1950). "Morphology and Taxonomy of Fungi." McGraw-Hill (Blakiston), New York.

Beug, H., Katz, F. E., and Gerisch, G. (1973a). Dynamics of antigenic membrane sites relating to cell aggregation in Dictyostelium discoideum. J. Cell Biol. 56, 647–658.

Beug, H., Katz, F. E., Stein, A., and Gerisch, G. (1973b). Quantitation of membrane sites in aggregating Dictyostelium cells by use of tritiated univalent antibody. Proc. Nat. Acad. Sci. U.S. 70, 3150–3154.

Blaskovics, J. S., and Raper, K. B. (1957). Encystment stages of Dictyostelium. Biol. Bull. 113, 58–88.

Bonner, J. T. (1944). A descriptive study of the development of the slime mold Dictyostelium discoideum. Amer. J. Bot. 31, 175–182.

Bonner, J. T. (1947). Evidence for the formation of cell aggregates by chemotaxis in the development of the slime mold Dictyostelium discoideum. J. Exp. Zool. 106, 1–26.

Bonner, J. T. (1949). The demonstration of acrasin in the later stages of the development of the slime mold Dictyostelium discoideum. J. Exp. Zool. 110, 259–271.

Bonner, J. T. (1952). The pattern of differentiation in amoeboid slime molds. Amer. Natur. 86, 79–89.

Bonner, J. T. (1957). A theory of the control of differentiation in the cellular slime molds. Quart. Rev. Biol. 32, 232–246.

Bonner, J. T. (1967). "The Cellular Slime Molds," 2nd ed. Princeton Univ. Press, Princeton, New Jersey.

Bonner, J. T. (1969). Hormones in social amoebae and mammals. Sci. Amer. 220, 78–92.

Bonner, J. T. (1970). Induction of stalk cell differentiation by cyclic AMP in the cellular slime mold Dictyostelium discoideum. Proc. Nat. Acad. Sci. U.S. 65, 110–113.

Bonner, J. T. (1971). Aggregation and differentiation in the cellular slime molds. Annu. Rev. Microbiol. 25, 75–92.

Bonner, J. T., and Adams, M. S. (1958). Cell mixtures of different species and strains of cellular slime molds. J. Embryol. Exp. Morphol. 6, 346–356.

Bonner, J. T., and Dodd, M. R. (1962a). Evidence for gas-induced orientation in the cellular slime molds. Develop. Biol. 5, 344–361.

Bonner, J. T., and Dodd, M. R. (1962b). Aggregation territories in the cellular slime molds. Biol. Bull. 122, 13–24.

Bonner, J. T., and Frascella, E. B. (1952). Mitotic activity in relation to differentiation in the slime mold Dictyostelium discoideum. J. Exp. Zool. 121, 561–571.

Bonner, J. T., and Frascella, E. B. (1953). Variations in cell size during the development of the slime mold, Dictyostelium discoideum. Biol. Bull. 104, 297–300.

Bonner, J. T., Clarke, W. W., Jr., Neely, C. L., Jr., and Slifkin, M. (1950). The orientation to light and the extremely sensitive orientation to temperature gradients in the slime mold Dictyostelium discoideum. J. Cell. Comp. Physiol. 36, 149–158.

Bonner, J. T., Chiquoine, A. D., and Kolderie, M. Q. (1955). A histochemical study of differentiation in the cellular slime molds. J. Exp. Zool. 130, 133–158.

Bonner, J. T., Hall, E. M., Noller, S., Oleson, F. B., Jr., and Roberts, A. B. (1972). Synthesis of cyclic AMP and phosphodiesterase in various species of cellular slime molds and its bearing on chemotaxis and differentiation. Develop. Biol. 29, 402–409.

Bovee, E. C. and Jahn, T. L. (1965). Mechanisms of movement in taxonomy of Sarcodina. II. The organization of subclasses and orders in relationship to the classes Autotractea and Hydraulea. *Amer. Midl. Natur.* **73**, 293–298.

Braselton, J. P., and Miller, C. E. (1973). Centrioles in *Sorosphaera. Mycologia* **65**, 220–226.

Brefeld, O. (1869). *Dictyostelium mucoroides*. Ein neuer Organismus aus der verwandtschaft der Myxomyceten. *Abh. Senckenberg. Naturforsch. Ges.* **7**, 85–107.

Brefeld, O. (1884). *Polysphondylium violaceum* und *Dictyostelium mucoroides* nebst Bemerkungen zur Systematik der Schleimpilze. *Untersuch. Gesammtgeb. Mykol.* **6**, 1–34.

Brewer, E., and Rusch, H. P. (1968). Effect of elevated temperature shocks on mitosis and on the initiation of DNA replication in *Physarum polycephalum. Exp. Cell Res.* **49**, 79–86.

Brown, R. M., Jr. (1969). Observations on the relationship of the Golgi apparatus to wall formation in the marine chrysophycean alga, *Pleurochrysis scherffelii* Pringsheim. *J. Cell Biol.* **41**, 109–123.

Brown, R. M., Jr., Franke, W. W., Kleinig, H., and Sitte, P. (1970). Scale formation in chrysophycean algae. I. Cellulosic and non-cellulosic wall components made by the Golgi apparatus. *J. Cell Biol.* **45**, 246–271.

Camp, W. G. (1937). The structure and activities of the myxomycete plasmodia. *Bull. Torrey Bot. Club* **64**, 307–335.

Carlile, M. J. (1972). The lethal interaction following plasmodial fusion between two strains of the myxomycete *Physarum polycephalum. J. Gen. Microbiol.* **71**, 581–590.

Carlile, M. J. (1973). Cell fusion and somatic incompatibility in myxomycetes. *Ber. Deut. Bot. Ges.* **86**, 123–139 and 177–178.

Carlile, M. J., and Dee, J. (1967). Plasmodial fusion and lethal interaction between strains in a myxomycete. *Nature (London)* **215**, 832–834.

Carroll, G., and Dykstra, R. (1966). Synaptinemal complexes in *Didymium iridis. Mycologia* **58**, 166–169.

Cavender, J. C. (1970). *Dictyostelium dimigraformum, Dictyostelium laterosorum,* and *Acytostelium ellipticum*: New Acrasieae from the American tropics. *J. Gen. Microbiol.* **62**, 113–123.

Cavender, J. C. (1973). Geographical distribution of Acrasieae. *Mycologia* **65**, 1044–1054.

Cavender, J. C., and Raper, K. B. (1965). The Acrasieae in nature. I. Isolation. II. Forest soil as a primary habitat. III. Occurrence and distribution in forests of eastern North America. *Amer. J. Bot.* **52**, 294–308.

Cavender, J. C., and Raper, K. B. (1968). The occurrence and distribution of Acrasieae in forests of subtropical and tropical America. *Amer. J. Bot.* **55**, 504–513.

Chadefaud, M. (1956). Sur un *Labyrinthula* de Roscoff. *C. R. Acad. Sci.* **243**, 1794–1797.

Charvat, I., Ross, I. K., and Cronshaw, J. (1973a). Ultrastructure of the plasmodial slime mold *Perichaena vermicularis*. I. Plasmodium. *Protoplasma* **76**, 333–351.

Charvat, I., Ross, I. K., and Cronshaw, J. (1973b). Ultrastructure of the plasmodial slime mold *Perichaena vermicularis*. II. Formation of the peridium. *Protoplasma* **78**, 1–19.

Chet, I. (1973). Changes in ribonucleic acid during differentiation of *Physarum polycephalum. Ber. Deut. Bot. Ges.* **86**, 77–92 and 174–175.

Chin, B., Friedrich, P. D., and Bernstein, I. A. (1972). Stimulation of mitosis in the myxomycete *Physarum polycephalum. J. Gen. Microbiol.* **71**, 93–101.

Cienkowsky, L. (1867). Ueber den Bau und die Entwickelung der Labyrinthuleen. *Arch. Mikrosk. Anat.* **3**, 274–310.

Cienkowsky, L. (1873). *Guttulina rosea.* Trans. Bot. Sect. 4th Meeting Russian Naturalists at Kazan.

Cienkowsky, L. (1876). Ueber einige Rhizopoden und verwandte Organismen. *Arch. Mikrosk. Anat.* **12**, 15–50.

Clark, J., and Collins, O. R. (1973a). Further studies on the genetics of plasmodial incompatibility in a Honduran isolate of *Didymium iridis. Mycologia* **65**, 507–518.

Clark, J., and Collins, O. R. (1973b). Directional cytotoxic reactions between incompatible plasmodia of *Didymium iridis. Genetics* **73**, 247–257.

Clark, M. A., Francis, D., and Eisenberg, R. (1973). Mating types in cellular slime molds. *Biochem. Biophys. Res. Commun.* **52**, 672–678.

Clegg, J. S., and Filosa, M. F. (1961). Trehalose in the cellular slime mould *Dictyostelium mucoroides. Nature* (*London*) **192**, 1077–1078.

Cohen, A. L. (1965). Slime molds. *In* "Encyclopaedia Britannica," Vol. 20, pp. 656–660.

Collins, O. R. (1961). Heterothallism and homothallism in two myxomycetes. *Amer. J. Bot.* **48**, 674–683.

Collins, O. R. (1963). Multiple alleles at the incompatibility locus in the myxomycete *Didymium iridis. Amer. J. Bot.* **50**, 477–480.

Collins, O. R. (1965). Evidence for a mutation at the incompatibility locus in the slime mold *Didymium iridis. Mycologia* **57**, 314–315.

Collins, O. R. (1966). Plasmodial compatibility in heterothallic and homothallic isolates of *Didymium iridis. Mycologia* **58**, 362–372.

Collins, O. R. (1969a). Experiments on the genetics of a slime mold, *Didymium iridis. Amer. Biol. Teach.* **31**, 33–36.

Collins, O. R. (1969b). Complementation between two color mutants in a true slime mold, *Didymium iridis. Genetics* **63**, 93–102.

Collins, O. R. (1972). Plasmodial fusion in *Physarum polycephalum:* Genetic analysis of a Turtox strain. *Mycologia* **64**, 1130–1137.

Collins, O. R., and Clark, J. (1966). Inheritance of the brown plasmodial pigment in *Didymium iridis, Mycologia* **58**, 743–751.

Collins, O. R., and Clark, J. (1968). Genetics of plasmodial compatibility and heterokaryosis in *Didymium iridis. Mycologia* **60**, 90–103.

Collins, O. R., and Erlebacher, B. (1969). Effects of two mutations on production of a red plasmodial pigment in the myxomycete *Didymium iridis. Can. J. Microbiol.* **15**, 1245–1247.

Collins, O. R., and Haskins, E. F. (1970). Evidence for polygenic control of plasmodial fusion in *Physarum polycephalum. Nature* (*London*) **226**, 279–280.

Collins, O. R., and Haskins, E. F. (1972). Genetics of somatic fusion in *Physarum polycephalum:* The PpII strain. *Genetics* **71**, 63–71.

Collins, O. R., and Ling, H. (1964). Further studies in multiple allelomorph heterothallism in the myxomycete *Didymium iridis. Amer. J. Bot.* **51**, 315–317.

Collins, O. R., and Ling, H. (1968). Clonally-produced plasmodia in heterothallic isolates of *Didymium iridis. Mycologia* **60**, 858–868.

Collins, O. R., and Ling, H. (1972). Genetics of somatic cell fusion in two isolates of *Didymium iridis. Amer. J. Bot.* **59**, 337–340.

Collins, O. R., and Tang, H-C. (1973). *Physarum polycephalum:* pH and plasmodium formation. *Mycologia* **65**, 232–236.

Collins, O. R., Alexopoulos, C. J., and Henney, M. (1964). Heterothallism in three isolates of the slime mold *Physarum pusillum. Amer. J. Bot.* 51, 679 (abstr.).

Cook, W. I. (1928). The methods of nuclear division in the Plasmodiophorales. *Ann. Bot. (London)* 42, 347–377.

Cook, W. I. (1933). A monograph of the Plasmodiophorales. *Arch. Protistenk.* 80, 179–254.

Copeland, H. F. (1956). "The Classification of Lower Organisms." Pacific Books, Palo Alto, California.

Cotter, D. A., Miura-Santo, L., and Hohl, H. R. (1969). Ultrastructural changes during germination of *Dictyostelium discoideum* spores. *J. Bacteriol.* 100, 1020–1026.

Coukell, M. B., and Walker, I. O. (1973). The basic nuclear proteins of the cellular slime mold *Dictyostelium discoideum. Cell Differentiation* 2, 87–95.

Cox, B. A., and Mackin, J. G. (1974). Studies on a new species of *Labyrinthula* (Labyrinthulales) isolated from the marine gastropod *Thais haemastoma floridana. Trans. Amer. Microsc. Soc.* 93, 62–70.

Cronshaw, J., and Charvat, I. (1973). Localization of B-glycerophosphatase activity in the myxomycete *Perichaena vermicularis. Can. J. Bot.* 51, 97–101.

Cummins, J. E. (1969). Nuclear DNA replication and transcription during the cell cycle of *Physarum. In* "The Cell Cycle" (G. M. Padilla, I. L. Cameron, and G. L. Whitson, eds.), pp. 141–158. Academic Press, New York.

Cummins, J. E., and Rusch, H. P. (1968). Natural synchrony in a slime mold. *Endeavour* 27, 124–129.

Dangeard, P. A. (1910). Etudes sur le dévelopment et la structure des organisms inférieurs. *Botaniste* 11, 1–332.

Dangeard, P. A. (1932). Observations sur la famille des Labyrinthulées et sur quelques autres parasites de *Cladophora. Botaniste* 24, 217–259.

Daniel, J. W. (1966). Light-induced synchronous sporulation of a myxomycete—the relation of initial metabolic changes to the establishment of a new cell state. *In* "Cell Synchrony" (I. L. Cameron and G. M. Padilla, eds.), pp. 117–152. Academic Press, New York.

Daniel, J. W., and Baldwin, H. (1964). Methods of culture for plasmodial Myxomycetes. *Methods Cell Physiol.* 1, 9–41.

Daniel, J. W., and Rusch, H. P. (1961). The pure culture of *Physarum polycephalum* on a partially defined medium. *J. Gen. Microbiol.* 25, 47–59.

Daniel, J. W., and Rusch, H. P. (1962). Method for inducing sporulation of pure cultures of the myxomycete *Physarum polycephalum. J. Bacteriol.* 83, 234–240.

Daniel, J. W., Kelley, J., and Rusch, H. P. (1962). Hematin-requiring plasmodial myxomycete. *J. Bacteriol.* 84, 1104–1110.

Daniel, J. W., Babcock, K. L., Sievert, A. H., and Rusch, H. P. (1963). Organic requirements and synthetic media for growth of the myxomycete, *Physarum polycephalum. J. Bacteriol.* 86, 324–331.

Darley, W. M., and Fuller, M. S. (1970). Cell wall chemistry and taxonomic position of *Schizochytrium. Amer. J. Bot.* 57, 761 (abstr.).

Darley, W. M., Porter, D., and Fuller, M. S. (1973). Cell wall composition and synthesis via Golgi-directed scale formation in the marine eucaryote, *Schizochytrium*, with a note on *Thraustochytrium sp. Arch. Mikrobiol.* 90, 89–106.

De Bary, A. (1859). "Die Mycetozoen. Ein Beitrag zur Kenntnis der Niedersten Organismen," 2nd ed. Engelmann, Leipzig.

De Bary, A. (1887). "Comparative Morphology and Biology of the Fungi, Mycetozoa and Bacteria" (English translation of the German original of 1884). Oxford Univ. Press (Clarendon), London and New York.

de Bruyne, C. (1890). Monadines et Chytridiacées, parasites des algues du Golfe de Naples. *Arch. Biol.* **10**, 43–104.

Dee, J. (1966). Multiple alleles and other factors affecting plasmodium formation in the true slime mold *Physarum polycephalum* Schw. *J. Protozool.* **13**, 610–616.

Dee, J. (1973). Aims and techniques of genetic analysis in *Physarum polycephalum. Ber. Deut. Bot. Ges.* **86**, 93–121 and 175–176.

Duboscq, O. (1921). *Labyrinthomyxa sauvageaui* n. g. n. sp., proteomyxée parasite de *Laminaria lejolisii* Sauvageau. *C. R. Soc. Biol.* **84**, 27–33.

Dutta, S. K., and Garber, E. D. (1961). The identification of physiological races of a fungal phytopathogen using strains of the slime mold *Acrasis rosea. Proc. Nat. Acad. Sci. U.S.* **47**, 990–993.

Dutta, S. K., and Mandel, M. (1972). Deoxyribonucleic acid base composition of some cellular slime molds. *J. Protozoology* **19**, 538–540.

Duval, J.-C. (1970). Bactéries d'allure rickettsienne parasites d'un myxomycète. 1. Cycle de l'infection et obtention d'une souche saine. *J. Microsc. (Paris)* **9**, 185–200.

Elliott, E. W. (1949). The swarm-cells of myxomycetes. *Mycologia* **41**, 141–170.

Ellis, T., Scheetz, R. W., and Alexopoulos, C. J. (1973). Ultrastructural observations on capillitial types in the Trichiales. *Trans. Amer. Microsc. Soc.* **92**, 65–79.

Ennis, H. L., and Sussman, M. (1958). The initiator cell for slime mold aggregation. *Proc. Nat. Acad. Sci. U.S.* **44**, 401–411.

Erbisch, F. H. (1964). Myxomycete spore longevity. *Mich. Bot.* **3**, 120–121.

Erdos, G. W. (1971). Ultrastructure and development of *Endemosarca*, a new plasmodial parasite of ciliates. Ph.D. Thesis, University of North Carolina, Chapel Hill.

Erdos, G. W. (1972a). The nuclear cycle and spore discharge in *Endemosarca hypsalyxis. Mycologia* **64**, 423–426.

Erdos, G. W. (1972b). Fine structure of the plasmodium of *Endemosarca. Cytobiologie* **5**, 83–100.

Erdos, G. W. (1973). A new species of *Endemosarca* from the Seychelles. *Mycologia* **65**, 229–232.

Erdos, G. W., and Olive, L. S. (1971). *Endemosarca*: A new genus with proteomyxid affinities. *Mycologia* **63**, 877–883.

Erdos, G. W., Nickerson, A., and Raper, K. B. (1972). Fine structure of macrocysts in *Polysphondylium violaceum. Cytobiologie* **6**, 351–366.

Erdos, G. W., Raper, K. B., and Vogen, L. (1973a). Mating types and macrocyst formation in *Dictyostelium discoideum. Proc. Nat. Acad. Sci. U.S.* **70**, 1828–1830.

Erdos, G. W., Nickerson, A., and Raper, K. B. (1973b). The fine structure of macrocyst germination in *Dictyostelium mucoroides. Develop. Biol.* **70**, 1828–1830.

Evans, T., and Suskind, D. (1971). Characterization of the mitochondrial DNA of the slime mold *Physarum polycephalum. Biochim. Biophys. Acta* **228**, 350–364.

Famintzin, A., and Woronin, M. (1873). Über zwei neue Formen von Schleimpilzen: *Ceratium hydnoides* und *Ceratium poroides. Mém. Acad. Imp. Sci. St. Petersburg, Ser.* **7**, **20**, 1–16.

Fayod, V. (1883). Beitrag zur Kenntnis niederer Myxomyceten. *Bot. Zeit.* **41**, 169–177.

Fell, H. B., and Hughes, A. F. (1949). Mitosis in the mouse: A study of living and fixed cells in tissue culture. *Quart. J. Microscop. Sci.* **90**, 355–381.

Filosa, M.. F. (1960). The effects of ethionine on the morphogenesis of cellular slime molds. *Anat. Rec.* **138**, 348. (abstr.).

Filosa, M. F., and Dengler, R. E. (1972). Ultrastructure of macrocyst formation in the cellular slime mold, *Dictyostelium mucoroides:* Extensive phagocytosis of amoebae by a specialized cell. *Develop. Biol.* **29**, 1–16.

Firtel, R. A. (1972). Changes in the expression of single-copy DNA during development of the cellular slime mold *Dictyostelium discoideum. J. Mol. Biol.* **66**, 363–377.

Firtel, R. A., and Bonner, J. (1972). Characterization of the genome of the cellular slime mold *Dictyostelium discoideum. J. Mol. Biol.* **66**, 339–361.

Firtel, R. A., Jacobson, A., and Lodish, H. F. (1972). Isolation and hybridization kinetics of messenger RNA from *Dictyostelium discoideum. Nature (London), New Biol.* **239**, 225–228.

Francis, D. W. (1962). Movement of pseudoplasmodia of *Dictyostelium discoideum.* Ph.D. Thesis, Univ. Wisconsin, Madison.

Francis, D. W. (1964). Some studies on phototaxis of *Dictyostelium. J. Cell Comp. Physiol.* **64**, 131–138.

Francis, D. W. (1965). Acrasin and the development of *Polysphondylium pallidum. Develop. Biol.* **12**, 329–346.

Franke, R. G., and Berry, J. A. (1972). Taxonomic application of isozyme patterns produced with disc electrophoresis of some myxomycetes, order Physarales. *Mycologia* **64**, 830–840.

Fukui, Y., and Takeuchi, I. (1971). Drug resistant mutants and appearance of heterozygotes in the cellular slime mould *Dictyostelium discoideum. J. Gen. Microbiol.* **67**, 307–317.

Fuller, M. S., and Rakatansky, R. (1966). A preliminary study of the carotenoids in *Acrasis rosea. Can. J. Bot.* **44**, 269–274.

Furtado, J. S., and Olive, L. S. (1970a). Ultrastructural studies of protostelids: The amoebo-flagellate stage. *Cytobiologie* **2**, 200–219.

Furtado, J. S., and Olive, L. S. (1970b). Ultrastructural studies of protostelids: The cyst stage of *Cavostelium bisporum. Protoplasma* **70**, 379–387.

Furtado, J. S., and Olive, L. S. (1971a). Ultrastructure of the protostelid *Ceratiomyxella tahitiensis,* including scale formation. *Nova Hedwigia* **21**, 537–576.

Furtado, J. S., and Olive, L. S. (1971b). Ultrastructural evidence of meiosis in *Ceratiomyxa fruticulosa. Mycologia* **63**, 413–416.

Furtado, J. S., and Olive, L. S. (1972). Scale formation in a primitive mycetozoan. *Trans. Amer. Microsc. Soc.* **91**, 594–596.

Furtado, J. S., Olive, L. S., and Jones, S. B., Jr. (1971). Ultrastructural studies of protostelids: The fruiting stage of *Cavostelium bisporum. Mycologia* **63**, 132–143.

Gaertner, A. (1964). Elektronenmikroskopische Untersuchungen zur Struktur der Geisseln von *Thraustochytrium* spec. *Veroeff. Inst. Meeresforsch. Bremerhaven* **9**, 25–28.

Gaertner, A. (1972). Characters used in the classification of thraustochytriaceous fungi. *Veroeff. Inst. Meeresforsch. Bremerhaven* **13**, 183–194.

Garber, E. D., and Dutta, S. K. (1962). Evidence for mixed cytoplasm in heterokaryons of *Collectotrichum lagenarium. Science* **135**, 665–666.

Garrod, D., and Ashworth, J. M. (1973). Development of the cellular slime mould *Dictyostelium discoideum.* In "Microbial Differentiation" (J. M. Ashworth and J. E. Smith, eds.), pp. 407–435. Cambridge Univ. Press, London and New York.

George, R. P. (1968). Cell organization and ultrastructure during the culmination of cellular slime molds. Ph.D. Thesis, University of Hawaii, Honolulu.

George, R. P., Hohl, H. R., and Raper, K. B. (1972). Ultrastructural development of stalk-producing cells in *Dictyostelium discoideum*, a cellular slime mould. *J. Gen. Microbiol.* **70**, 477–489.

Gerisch, G. (1961). Zellfunktionen und Zellfunktionswechsel in der Entwicklung von *Dictyostelium discoideum*. II. Aggregation homogener Zellpopulationen und Zentrenbildung. *Develop. Biol.* **3**, 685–724.

Gerisch, G. (1963). Eine für *Dictyostelium* ungewöhnliche Aggregationsweise. *Naturwissenschaften* **50**, 160–161.

Gerisch, G., Malchow, D., Wilhelms, H., and Lüderiz, O. (1969). Artspezifität polysaccharid-haltiger Zellmembran-Antigene von *D. discoideum*. *Eur. J. Biochem.* **9**, 229–236.

Gerisch, G., Malchow, D., Riedel, V., Müller, E., and Every, M. (1972). Cyclic AMP phosphodiesterase and its inhibitor in slime mould development. *Nature* (*London*), *New Biol.* **235**, 90–92.

Gezelius, K. (1959). The ultrastructure of cells and cellulose membranes in Acrasiae. *Exp. Cell Res.* **18**, 425–453.

Gilbert, H. C. (1935). Critical events in the life history of *Ceratiomyxa. Amer. J. Bot.* **22**, 52–74.

Gilbert, H. C., and Martin, G. W. (1933). Myxomycetes found on the bark of living trees. *Stud. Natur. Hist. Iowa Univ.* **15**, 3–8.

Goldstein, S. (1963a). Morphological variation and nutrition of a new monocentric marine fungus. *Arch. Mikrobiol.* **45**, 101–110.

Goldstein, S. (1963b). Studies of a new species of *Thraustochytrium* that displays light stimulated growth. *Mycologia* **55**, 799–811.

Goldstein, S. (1963c). Development and nutrition of new species of *Thraustochytrium. Amer. J. Bot.* **50**, 271–279.

Goldstein, S. (1973). Zoosporic marine fungi (Thraustochytriaceae and Dermatocystidiaceae). *Annu. Rev. Microbiol.* **27**, 13–26.

Goldstein, S., and Belsky, M. (1964). Axenic culture of a new marine phycomycete possessing an unusual type of asexual reproduction. *Amer. J. Bot.* **51**, 72–78.

Goldstein, S., Moriber, L., and Hershenov, B. (1964). Ultrastructure of *Thraustochytrium aureum*, a biflagellate marine phycomycete. *Mycologia* **56**, 897–904.

Goldstone, E. Banerjee, S. D., Allen, J. R., Lee, J. J., Hutner, S. H., Bacchi, C. J., and Melville, J. F. (1966). Minimal defined media for vegetative growth of the acrasian *Polysphondylium pallidum* WS-320. *J. Protozool.* **13**, 171–174.

Goodman, E. M. (1972). Axenic culture of myxamoebae of the myxomycete *Physarum polycephalum. J. Bacteriol.* **111**, 242–247.

Goodman, E. M., and Rusch, H. P. (1970). Ultrastructural changes during spherule formation in *Physarum polycephalum. J. Ultrastruct. Res.* **30**, 172–183.

Goodwin, D. C. (1961). Morphogenesis of the sporangium of *Comatricha. Amer. J. Bot.* **48**, 148–154.

Gray, W. D. (1938). The effect of light on the fruiting of Myxomycetes. *Amer. J. Bot.* **25**, 511–522.

Gray, W. D. (1939). The relation of pH and temperature to the fruiting of *Physarum polycephalum. Amer. J. Bot.* **26**, 709–714.

Gray, W. D., and Alexopoulos, C. J. (1968). "Biology of the Myxomycetes." Ronald Press, New York.

Gregg, J. H. (1956). Serological investigations of cell adhesion in the slime molds,

Dictyostelium discoideum, D. purpureum, and *Polysphondylium violaceum. J. Gen. Physiol.* **39,** 813–820.

Gregg, J. H. (1961). An immunoelectrophoretic study of the slime mold, *Dictyostelium discoideum. Develop. Biol.* **3,** 757–766.

Gregg, J. H. (1964). Developmental processes in cellular slime molds. *Phys. Rev.* **44,** 631–656.

Gregg, J. H. (1965). Regulation in the cellular slime molds. *Develop. Biol.* **12,** 377–393.

Gregg, J. H. (1966). Organization and synthesis in the cellular slime molds. *In* "The Fungi" (G. C. Ainsworth and A. S. Sussman, eds.), Vol. 2, pp. 235–281. Academic Press, New York.

Gregg, J. H. (1971). Developmental potential of isolated *Dictyostelium myxamoebae. Develop. Biol.* **26,** 478–485.

Gregg, J. H., and Badman, W. S. (1970). Morphogenesis and ultrastructure in *Dictyostelium. Develop. Biol.* **22,** 96–111.

Gregg, J. H., and Nesom, M. (1973). Response of *Dictyostelium* plasma membranes to adenosine 3′:5′-cyclic monophosphate. *Proc. Nat. Acad. Sci. U.S.* **70,** 1630–1633.

Guttes, E., and Guttes, S. (1960). Pinocytosis in the myxomycete *Physarum polycephalum. Exp. Cell Res.* **20,** 239–241.

Guttes, E., and Guttes, S. (1964). Thymidine incorporation by mitochondria in *Physarum polycephalum. Science* **145,** 1057–1058.

Guttes, E., Guttes, S., and Rusch, H. P. (1959). Synchronization of mitosis by the fusion of the plasmodia of *Physarum polycephalum* grown in pure culture. *Develop. Biol.* **3,** 588–614.

Guttes, E., Guttes, S., and Rusch, H. P. (1961). Morphological observations on growth and differentiation of *Physarum polycephalum* grown in pure culture. *Develop. Biol.* **3,** 588–614.

Guttes, S., Guttes, E., and Hadek, R. (1966). Occurrence and morphology of a fibrous body in the mitochondria of the slime mold *Physarum polycephalum. Experientia* **22,** 452–454.

Guttes, S., Guttes, E., and Ellis, R. A. (1968). Electron microscope study of mitosis in *Physarum polycephalum. J. Ultrastruct. Res.* **22,** 508–528.

Haeckel, E. (1868). Monographie der Moneren. *Jena. Z. Med. Naturwiss.* **4,** 64–137.

Hagiwara, H. (1971). The Acrasiales of Japan. I. *Natur. Sci. & Mus., Tokyo, Bull.* **14,** 351–366.

Hagiwara, H. (1972). Acrasiales. *Natur. Sci. & Mus., Tokyo, Mem.* No. 5, pp. 173–177.

Hagiwara, H. (1973a). Enumeration of the Dictyosteliaceae. *Natur. Sci. & Mus. Tokyo, Bull.* **16,** 493–496.

Hagiwara, H. (1973b). The Acrasiales of Japan. II. *Rep. Tottori Mycol. Inst.* (*Jap.*) No. **10,** 591–595.

Harper, R. A., and Dodge, B. O. (1914). The formation of capillitium in certain myxomycetes. *Ann. Bot.* (*London*) **28,** 1–18.

Haskins, E. F. (1968). Developmental studies on the true slime mold *Echinostelium minutum. Can. J. Microbiol.* **14,** 1309–1315.

Haskins, E. F. (1970). Axenic culture of myxamoebae of the myxomycete *Echinostelium minutum. Can. J. Bot.* **48,** 663–664.

Haskins, E. F., Hinchee, A. A., and Cloney, R. A. (1971). The occurrence of synaptonemal complexes in the slime mold *Echinostelium minutum* de Bary. *J. Cell Biol.* **51**, 898–903.

Hatano, S., and Oosawa, F. (1962). Actin-like protein of myxomycete plasmodium. I. Extraction and cross reaction with myosin from muscle. *Annu. Rep. Res. Group Biophys. Jap.* **2**, 29–30.

Hatano, S., and Oosawa, F. (1964). Actin-like protein of myxomycete plasmodium. IV. Purification and observations of some physico-chemical properties. *Annu. Rep. Res. Group Biophys. Jap.* **4**, 25–28.

Harrison, J. L., and Jones, E. B. G. (1974). Ultrastructural aspects of the marine fungus *Japonochytrium* sp. *Arch. Microbiol.* **96**, 305–317.

Haugli, F. B., and Dove, W. F. (1972a). Genetics and biochemistry of cycloheximide resistance in *Physarum polycephalum*. *Mol. Gen. Genet.* **118**, 97–107.

Haugli, F. B., and Dove, W. F. (1972b). Mutagenesis and mutant selection in *Physarum polycephalum*. *Mol. Gen. Genet.* **118**, 109–124.

Hemmes, D. E., Kojima-Buddenhagen, E. S., and Hohl, H. H. (1972). Structural and enzymatic analysis of the spore wall layers in *Dictyostelium discoideum*. *J. Ultrastruct. Res.* **41**, 406–417.

Henney, H. R., and Henney, M. (1968). Nutritional requirements for the growth in pure culture of the myxomycete *Physarum rigidum* and related species. *J. Gen. Microbiol.* **53**, 333–339.

Henney, H. R., and Lynch, T. (1969). Growth of *Physarum flavicomum* and *Physarum rigidum* in chemically defined minimal media. *J. Bacteriol.* **99**, 531–534.

Henney, M., and Henney, H. R. (1968). The mating-type systems of the myxomycetes *Physarum rigidum* and *P. flavicomum*. *J. Gen. Microbiol.* **53**, 321–332.

Hohl, H. R. (1966). The fine structure of the slimeways in *Labyrinthula*. *J. Protozool.* **13**, 41–43.

Hohl, H. R., and Hamamoto, S. (1968). Lamellate structures in the nucleolus of the cellular slime mold *Acrasis rosea*. *Pac. Sci.* **22**, 402–407.

Hohl, H. R., and Hamamoto, S. (1969a). Ultrastructure of spore differentiation in *Dictyostelium:* The prespore vacuole. *J. Ultrastruct. Res.* **26**, 442–453.

Hohl, H. R., and Hamamoto, S. (1969b). Ultrastructure of *Acrasis rosea*, a cellular slime mold, during development. *J. Protozool.* **16**, 333–344.

Hohl, H. R., and Jehli, J. (1973). The presence of cellulose microfibrils in the proteinaceous slime track of *Dictyostelium discoideum*. *Arch. Mikrobiol.* **92**, 178–187.

Hohl, H. R., and Raper, K. B. (1963). Nutrition of cellular slime molds. II. Growth of *Polysphondylium pallidum* in axenic culture. *J. Bacteriol.* **85**, 199–206.

Hohl, H. R., Hamamoto, S., and Hemmes, D. E. (1968). Ultrastructural aspects of cell elongation, cellulose synthesis, and spore differentiation in *Acytostelium leptosomum*, a cellular slime mold. *Amer. J. Bot.* **55**, 783–796.

Hohl, H. R., Miura-Santo, L. Y., and Cotter, D. A. (1970). Ultrastructural changes during formation and germination of microcysts in *Polysphondylium pallidum*, a cellular slime mould. *J. Cell Sci.* **7**, 285–305.

Hollande, A., and Enjumet, M. (1955). Sur l'évolution et la systématique des Labyrinthulidae; étude de *Labyrinthula algeriensis* nov. sp. *Ann. Sci. Natur., Zool. Biol. Anim.* [11] **17**, 357–368.

Honigberg, B. M., Balamuth, W., Bovee, E. C., Corliss, J. O., Gojdics, M., Hall, R. P., Kudo, R. R., Levine, N. D., Loeblich, A. R., Weiser, J., and Wenrich, D. H. (1964). A revised classification of the phylum Protozoa. *J. Protozool.* **11**, 7–20.

Howard, F. L. (1931). The life history of *Physarum polycephalum*. *Amer. J. Bot.* **18**, 116–133.

Howard, F. L. (1932). Nuclear division in plasmodia of *Physarum*. *Ann. Bot. (London)* **46**, 461–477.

Huffman, D. M., and Olive, L. S. (1963). A significant morphogenetic variant of *Dictyostelium mucoroides*. *Mycologia* **55**, 333–344.

Huffman, D. M., and Olive, L. S. (1964). Engulfment and anastomosis in the cellular slime molds (Acrasiales). *Amer. J. Bot.* **51**, 465–471.

Huffman, D. M., Kahn, A. J., and Olive, L. S. (1962). Anastomosis and cell fusions in *Dictyostelium*. *Proc. Nat. Acad. Sci. U.S.* **48**, 1160–1164.

Hung, C.-Y., and Olive, L. S. (1972a). Ultrastructure of the spore wall in *Echinostelium*. *Mycologia* **64**, 1160–1163.

Hung, C.-Y., and Olive, L. S. (1972b) Ultrastructure of the amoeboid cell and its vacuolar system in *Protosteliopsis fimicola*. *Mycologia* **64**, 1312–1327.

Hung, C.-Y., and Olive, L. S. (1973a). Ultrastructure of the ameboid cells of *Protostelium zonatum* (Mycetozoa). *J. Protozool.* **20**, 252–263.

Hung, C.-Y., and Olive, L. S. (1973b). Intranuclear inclusions in the ameboid cells of *Protostelium zonatum*. *J. Protozool.* **20**, 263–267.

Hütterman, A. (1973). Biochemical events during spherule formation of *Physarum polycephalum*. *Ber. Deut. Bot. Ges.* **86**, 55–76 and 172–174.

Ikeda, T., and Takeuchi, I. (1971). Isolation and characterization of a prespore specific structure of the cellular slime mold, *Dictyostelium discoideum*. *Develop., Growth & Differentiation* **13**, 221–229.

Indira, P. U. (1964). Swarmer formation from plasmodia of myxomycetes. *Trans. Brit. Mycol. Soc.* **47**, 531–533.

Indira, P. U. (1969). In vitro cultivation of some myxomycetes. *Nova Hedwigia* **18**, 627–636.

Indira, P. U. (1971). The life-cycle of *Stemonitis herbatica*. II. *Trans. Brit. Mycol. Soc.* **56**, 251–259.

Indira, P. U., and Kalyanasundaram, R. (1963). Preliminary investigations in culture of some myxomycetes. *Ber. Schweiz. Bot. Ges.* **73**, 381–388.

Ing, B., and Nannenga-Bremekamp, N. W. (1967). Notes on Myxomycetes. XIII. *Stemphytocarpus* nov. gen. stemonitacearum. *Proc. Kon. Ned. Akad. Wetensch. Ser. C* **70**, 217–233.

Ingram, D. S. (1969). Growth of *Plasmodiophora brassicae* in host callus. *J. Gen. Microbiol.* **55**, 9–18.

Jahn, E. (1911). Myxomycetenstudien, 8. Der Sexualakt. *Ber. Deut. Bot. Ges.* **29**, 231–247.

Jahn, E. (1933). Myxomycetenstudien, 15. Somatische und generative Kernteilungen. *Ber. Deut. Bot. Ges.* **51**, 377–385.

Jahn, T. L. (1964). Protoplasmic flow in the mycetozoan, *Physarum*. II. The mechanism of flow; a reevaluation of the contraction-hydraulic theory and of the diffusion drag hypothesis. *Biorheology* **2**, 133–152.

Jahn, T. L., Rinaldi, R. A., and Brown, M. (1964). Protoplasmic flow in the mycetozoan, *Physarum*. I. Geometry of the plasmodium and the observable facts of flow. *Biorheology* **2**, 123–131.

Jockusch, B. M. (1973). Nuclear proteins in *Physarum polycephalum*. *Ber. Deut. Bot. Ges.* **86**, 39–54 and 171–172.

Jockusch, B. M., Brown, D. F., and Rusch, H. P. (1971). Synthesis and some properties of an actin-like nuclear protein in the slime mold *Physarum polycephalum. J. Bacteriol.* **108,** 705–714.

Jockusch, B. M., Ryser, U., and Behnke, O. (1973). Myosin-like protein in *Physarum* nuclei, *Exp. Cell Res.* **76,** 464–466.

Johnson, T. W., and Sparrow, F. K., Jr. (1961). "Fungi in Oceans and Estuaries." Cramer, New York.

Jones, E. B. G., and Alderman, D. J. (1972). *Althornia crouchii* gen. et sp. nov., a marine biflagellate fungus. *Nova Hedwigia* **21,** 381–400.

Jump, J. A. (1954). Studies on sclerotization in *Physarum polycephalum. Amer. J. Bot.* **41,** 561–567.

Kahn, A. J. (1964a). Some aspects of cell interaction in the development of the slime mold, *Dictyostelium purpureum. Develop. Biol.* **9,** 1–19.

Kahn, A. J. (1964b). The influence of light on cell aggregation in *Polysphondylium pallidum. Biol. Bull.* **127,** 85–96.

Kahn, A. J. (1968). An analysis of the spacing of aggregation centers in *Polysphondylium pallidum. Develop. Biol.* **18,** 149–162.

Kamiya, N. (1950). The protoplasmic flow in the myxomycete plasmodium as revealed by a volumetric analysis. *Protoplasma* **39,** 344–357.

Karling, J. S. (1942). "The Plasmodiophorales." Published by the Author.

Karling, J. S. (1944). *Phagomyxa algarum* n. gen., n. sp., an unusual parasite with plasmodiophoralan and proteomyxean characteristics. *Amer. J. Bot.* **31,** 38–52.

Karling, J. S. (1968). "The Plasmodiophorales," 2nd rev. ed. Hafner, New York.

Katz, E. R., and Bourguignon, L. (1974). The cell cycle and its relationship to aggregation in the cellular slime mold, *Dictyostelium discoideum. Develop. Biol.* **36,** 82–87.

Katz, E. R., and Sussman, M. (1972). Parasexual recombination in *Dictyostelium discoideum:* Selection of stable diploid heterozygotes and stable haploid segregants. *Proc. Nat. Acad. Sci. U.S.* **69,** 495–498.

Kazama, F. Y. (1972a). Ultrastructure of *Thraustochytrium* sp. zoospores. I. Kinetosomes. *Arch. Mikrobiol.* **83,** 179–188.

Kazama, F. Y. (1972b). Ultrastructure of *Thraustochytrium* sp. zoospores. II. Striated inclusions. *J. Ultrastruc. Res.* **41,** 60–66.

Kazama, F. Y. (1973). Ultrastructure of *Thraustochytrium* sp. zoospores. III. Cytolysomes and acid phosphatase distribution. *Arch. Mikrobiol.* **89,** 95–104.

Kazama, F. Y. (1974). Ultrastructure of *Thraustochytrium* sp. zoospores. IV. External morphology with notes on the zoospores of *Schizochytrium* sp. *Mycologia* **66,** 272–280.

Kazama, F. Y., and Aldrich, H. C. (1972). Digestion and the distribution of acid phosphatase in the myxamoebae of *Physarum flavicomum. Mycologia* **64,** 529–538.

Kazama, F. Y., and Schornstein, K. L. (1972). Herpes-like virus particles associated with a fungus. *Science* **177,** 696–697.

Kerr, N. S., and Kerr, S. J. (1967). A new hypothesis concerning plasmodium formation in the true slime mold, *Didymium nigripes. Amer. Zool.* **7,** 230.

Kerr, S. J. (1967). A comparative study of mitosis in amoebae and plasmodia of the true slime mold *Didymium nigripes. J. Protozool.* **14,** 439–445.

Kerr, S. J. (1968). Ploidy level in the true slime mould *Didymium nigripes. J. Gen. Microbiol.* **53,** 9–15.

Kerr, S. J. (1970). Nuclear size in plasmodia of the true slime mould *Didymium nigripes. J. Gen. Microbiol.* **63,** 347–356.

Keskin, B. (1964). *Polymyxa betae* n. sp., ein Parasit in den Wurzeln von *Beta vulgaris* Tournefort, besonders während der Jugendentwicklung der Zuckerrübe. *Arch. Mikrobiol.* **49**, 348–374.

Keskin, B. (1971). Beitrag zur Protomitose bei *Polymyxa betae* Keskin. *Arch. Mikrobiol.* **77**, 344–348.

Keskin, B., and Fuchs, W. H. (1969). Der Infektionsvorgang bei *Polymyxa betae*. *Arch. Mikrobiol.* **68**, 218–226.

Kirk, D., McKeen, W. E., and Smith, R. (1971). Cytoplasmic connections between *Dictyostelium discoideum* cells. *Can. J. Bot.* **49**, 19–20.

Kislev, N., and Chet, I. (1973). Scanning electron microscopy of sporulating cultures of the myxomycete *Physarum polycephalum*. *Tissue & Cell* **5**, 349–357.

Klie, H., and Mach, F. (1968). Licht- und elektronmikroscopische Untersuchungen über die Wirkung von Labyrinthula-Enzymen auf Bakterien und Hefezellen. *Z. Allg. Mikrobiol.* **8**, 385–395.

Kobayasi, Y., and Ookubo, M. (1953). Studies on the marine phycomycetes. *Natur. Sci. & Mus., Tokyo, Bull.* **33**, 53–65.

Koevenig, J. L. (1961). Slime molds. I. Life cycle. U5518 color film. Bureau Audio Visual Instruct., Ext. Div., University of Iowa, Iowa City.

Koevenig, J. L. (1964). Studies on life cycle of *Physarum gyrosum* and other myxomycetes. *Mycologia* **56**, 170–184.

Koevenig, J. L., and Jackson, R. C. (1966). Plasmodial mitoses and polyploidy in the myxomycete *Physarum polycephalum*. *Mycologia* **58**, 662–667.

Köle, A. P. (1954). A contribution to the knowledge of *Spongospora subterranea* (Wallr.) Lagerh., the cause of powdery scab of potatoes. *Tijdschr. Plantenziekten* **60**, 1–65.

Köle, A. P., and Gielink, A. J. (1961). Electron microscope observations on the flagella of the zoosporangial zoospores of *Plasmodiophora brassicae* and *Spongospora subterranea*. *Proc. Kon. Ned. Akad. Wetensch., Ser. C* **64**, 157–161.

Konijn, T. M. (1972a). Cyclic AMP as a first messenger. *Advan. Cyclic Nucleotide Res.* **1**, 17–31.

Konijn, T. M. (1972b). Cyclic AMP and cell aggregation in the cellular slime molds. *Acta Protozool.* **11**, 137–143.

Konijn, T. M., and Raper, K. B. (1961). Cell aggregation in *Dictyostelium discoideum*. *Develop. Biol.* **3**, 725–756.

Konijn, T. M., van de Meene, J. G. C., Bonner, J. T., and Barkley, D. S. (1967). The acrasin activity of adenosine-3',5'-cyclic phosphate. *Proc. Nat. Acad. Sci. U.S.* **58**, 1152–1154.

Konijn, T. M., Chang, Y-Y., and Bonner, J. T. (1969). Synthesis of cyclic AMP in *Dictyostelium discoideum* and *Polysphondylium pallidum*. *Nature (London)* **224**, 1211–1212.

Kudo, R. R. (1963). "Protozoology," 5th ed. Thomas, Springfield, Illinois.

Kuehn, G. D. (1971). An adenosine 3',5' monophosphate inhibited protein kinase from *Physarum polycephalum*. *J. Biol. Chem.* **246**, 6366–6369.

Kusaba, T., and Toyama, A. (1970). Transmission of soil-born barley mosaic virus. 1. Infectivity of diseased root washings. *Ann. Phytopathol. Soc. Jap.* **36**, 214–222.

Lankester, R. (1885). Protozoa. *In* "Encyclopaedia Britannica," Vol. 19, pp. 830–866.

Larpent, J.-P. (1972). Quelques données nouvelles de la biologie des Acrasiales et des Myxomycétales. *Botaniste* **55**, 39–69.

Ledingham, G. A. (1933). Studies on *Polymyxa graminis*, n. gen., n. sp., a plasmodiophoraceous root parasite of wheat. *Can J. Res., Sect. C* **17**, 38–51.

Ledingham, G. A. (1935). Occurrence of zoosporangia in *Spongospora subterranea* (Wallroth) Lagerheim. *Nature* (*London*) 135, 394.

LeStourgeon, W. M., Bohnstedt, C. F., and Thimell, P. (1971). Supportive evidence for postcleavage meiosis in *Physarum flavicomum*. *Mycologia* 63, 1002–1012.

Levine, N. D., and Corliss, J. O. (1963). Two new subclasses of sarcodines Labyrinthulia subcl. nov. and Proteomyxidia subcl. nov. *J. Protozool.* 10, Suppl., 27 (abstr.)

Lieth, H. (1954). Die Pigmente von *Didymium eunigripes* und ihre Beziehungen zur Lichtsabsorbtien. *Ber. Deut. Bot. Ges.* 67, 323–325.

Lieth, H. (1956). Die Wirkung des Grünlichtes auf die Fruchtkörperbildung bei *Didymium eunigripes*. *Arch. Mikrobiol.* 24, 91–104.

Ling, H. (1968). Light and fruiting in *Didymium iridis*. *Mycologia* 60, 966–970.

Ling, H. (1971). Genetics of somatic fusion in a myxomycete: F2 studies. *Protoplasma* 73, 407–416.

Ling, H., and Collins, O. R. (1970a). Control of plasmodial fusion in a Panamanian isolate of *Didymium iridis*. *Amer J. Bot.* 57, 292–298.

Ling, H., and Collins, O. R. (1970b). Linkage studies in the true slime mold *Didymium iridis*. *Amer. J. Bot.* 57, 299–303.

Lister, A. (1925). "A Monograph of the Mycetozoa" (revised by G. Lister), 3rd ed. Brit. Mus. Natur. Hist., London.

Locquin, M. (1949). Recherches sur les simblospores des Myxomycètes. *Bull. Soc. Linn. Lyon* 18, 43–46.

Locquin, M. (1967). Mycotaxia 1967. 1A. Myxomycetes, Genera. Locquin, 10 Rue Talma, Paris.

Loeblich, A. R., and Tappan, H. (1961). Suprageneric classification of the Rhizopoda. *J. Paleontol.* 35, 245–330.

Loewy, A. G. (1952). An actomyosin-like substance from the plasmodium of a myxomycete. *J. Cell. Comp. Physiol.* 40, 127–156.

Loomis, W. F. (1972). Role of the surface sheath in the control of morphogenesis in *Dictyostelium discoideum*. *Nature* (*London*), *New Biol.* 240, 6–9.

Loomis, W. F., and Ashworth, J. M. (1968). Plaque-size mutants of the cellular slime mould *Dictyostelium discoideum*. *J. Gen. Microbiol.* 53, 181–186.

Lucas, S., Bazin, M., and Kerr, N. (1968). Observations on the differentiation of plasmodia into fruiting bodies by the true slime mold, *Didymium nigripes*. *J. Gen. Microbiol.* 53, 17–21.

Lynch, T. J., and Henney, H. R., Jr. (1973). Carbohydrate metabolism during differentiation (sclerotization) of the myxomycete *Physarum flavicomum*. *Arch. Mikrobiol.* 90, 189–198.

Macbride, T. H., and Martin, G. W. (1934). "The Myxomycetes." Macmillan, New York.

McCormick, J. J., Blomquist, J., and Rusch, H. P. (1970a). Isolation and characterization of an extracellular polysaccharide from *Physarum polycephalum*. *J. Bacteriol.* 104, 1110–1118.

McCormick, J. J., Blomquist, J., and Rusch, H. P. (1970b). Isolation and characterization of a galactosamine wall from spores and spherules of *Physarum polycephalum*. *J. Bacteriol.* 104, 1119–1125.

Mackin, J. G., and Ray, S. M. (1966). The taxonomic relationships of *Dermocystidium marinum* Mackin, Owen, and Collier, *J. Invertebr. Pathol.* 8, 544–545.

McManus, M. A. (1958). *In vivo* studies of plasmogamy in *Ceratiomyxa*. *Bull. Torrey Bot. Club* 85, 28–37.

McManus, M. A. (1962). Some observations on plasmodia of the Trichiales. *Mycologia* **54**, 78–90.

McManus, M. A., and Roth, L. E. (1965). Fibrillar differentiation in myxomycete plasmodia. *J. Cell Biol.* **25**, 305–318.

McManus, M. A., and Ruch, L. (1965). Phase contrast and electron microscope observations on membranous organelles in myxomycete plasmodia. *Proc. Iowa Acad. Sci.* **72**, 40–43.

Maeda, Y. (1971a). Studies on a specific structure in differentiating slime mold cells. *Mem. Fac. Sci. Kyoto Univ., Ser. Biol.* **4**, 97–107.

Maeda, Y. (1971b). Formation of a prespore specific structure from a mitochondrion during development of the cellular slime mold *Dictyostelium discoideum*. *Develop., Growth & Differentiation* **13**, 211–219.

Maeda, Y., and Maeda, M. (1973). The calcium content of the cellular slime mold, *Dictyostelium discoideum*, during development and differentiation. *Exp. Cell Res.* **82**, 125–130.

Maeda, Y., and Takeuchi, I. (1969). Cell differentiation and fine structures in the development of the cellular slime molds. *Develop., Growth & Differentiation* **11**, 231–245.

Maeda, Y., Sugita, K., and Takeuchi, I. (1973). Fractionation of the differentiated types of cells constituting the pseudoplasmodia of the cellular slime molds. *Bot. Mag.* **86**, 5–12.

Maire, R., and Tison, A. (1911). Sur quelques Plasmodiophoracées non hypertrophiantes. *C. R. Acad. Sci.* **152**, 206–208.

Malchow, D., Nägele, B., Schwarz, H., and Gerisch, G. (1972). Membrane-bound cyclic AMP phosphodiesterase in chemotactically responding cells of *Dictyostelium discoideum*. *Eur. J. Biochem.* **28**, 136–142.

Manton, I. (1967). Further observations on the fine structure of *Chrysochromulina chiton* with special reference to the haptonema, "peculiar" Golgi structure and scale production. *J. Cell Sci.* **2**, 265–272.

Martin, G. W. (1960). The systematic position of the myxomycetes. *Mycologia* **52**, 119–129.

Martin, G. W., and Alexopoulos, C. J. (1969). "The Myxomycetes." Univ. of Iowa Press, Iowa City.

Mercer, E. H., and Shaffer, B. M. (1960). Electron microscopy of solitary and aggregated slime mold cells. *J. Biophys. Biochem. Cytol.* **7**, 353–356.

Miller, C. E. (1958). Morphology and cytology of the zoosporangia and cystosori of *Sorosphaera veronicae*. *J. Elisha Mitchell Sci. Soc.* **74**, 49–64.

Mims, C. W. (1971). An ultrastructural study of spore germination in the myxomycete *Arcyria cinerea*. *Mycologia* **63**, 586–601.

Mims, C. W. (1972a). An ultrastructural study of precleavage mitosis in the myxomycete *Arcyria cinerea*. *J. Gen. Microbiol.* **71**, 53–62.

Mims, C. W. (1972b). Centrioles and Golgi apparatus in postmeiotic spores of the myxomycete *Stemonitis virginiensis*. *Mycologia* **64**, 452–456.

Mims, C. W. (1973). A light and electron microscope study of sporulation in the myxomycete *Stemonitis virginiensis*. *Protoplasma* **77**, 35–54.

Mims, C. W., and Rogers, M. A. (1973). An ultrastructural study of spore germination in the myxomycete *Stemonitis virginiensis*. *Protoplasma* **78**, 243–254.

Mishou, K., and Haskins, E. F. (1971). A survey of the Acrasieae in the soils of Washington state. *Syesis* **4**, 179–184.

Moens, P. B., and Perkins, F. O. (1969). Chromosome number of a small protist: Accurate determination. Science 166, 1289–1291.

Mohberg, J., Babcock, K., Haugli, F. B., and Rusch, H. P. (1973). Nuclear DNA content and chromosome numbers in the myxomycete Physarum polycephalum. Develop. Biol. 34, 228–245.

Molisch, H. (1926). Pseudoplasmodium aurantiacum n. g. et n. sp. eine neue Acrasiee aus Japan. Sci. Rep. Tohoku Univ., Ser. 4 1, 119–134.

Mukherjee, K. L., and Zabka, G. G. (1964). Studies of multiple allelism in the myxomycete Didymium iridis. Can. J. Bot. 42, 1459–1466.

Murray, A. W., Spiszman, M., and Atkinson, D. E. (1971). Adenosine 3',5' monophosphate phosphodiesterase in the growth medium of Physarum polycephalum. Science 171, 496–498.

Nachmias V. (1972). Filament formation by purified Physarum myosin. Proc. Nat. Acad Sci. U.S. 69, 2011–2014.

Nachmias, V., and Ingram, W. C. (1970). Actomyosin from Physarum polycephalum: Electron microscopy of myosin-enriched preparations. Science 170, 743–745.

Nakajima, H. (1964). The mechanochemical system behind streaming in Physarum. In "Primitive Motile Systems in Cell Biology" (R. D. Allen and N. Kamiya, eds.), pp. 111–120. Academic Press, New York.

Nakajima, H., and Hatano, S. (1962). Acetylcholinesterase in the plasmodium of the myxomycete, Physarum polycephalum. J. Cell. Physiol. 59, 259–263.

Nannenga-Bremekamp, N. E. (1967). Notes on Myxomycetes. XII. A revision of the Stemonitales. Proc. Kon. Ned. Akad. Wetensch., Ser. C 70, 201–216.

Nelson, N., Olive, L. S., and Stoianovitch, C. (1967). A new species of Dictyostelium from Hawaii. Amer. J. Bot. 54, 354–358.

Nesom, M. (1973). Life cycle and ultrastructure of the cellular slime mold, Copromyxa arborescens. Ph.D. Thesis, University of North Carolina, Chapel Hill.

Nesom, M., and Olive, L. S. (1972). Copromyxa arborescens, a new cellular slime mold. Mycologia 64, 1359–1362.

Nestle, M., and Sussman, M. (1972). The effect of cyclic AMP on morphogenesis and enzyme accumulation in Dictyostelium discoideum. Develop. Biol. 28, 545–554.

Newell, P. C. (1971). The development of the cellular slime mould Dictyostelium discoideum: A model system for the study of cellular differentiation. Essays Biochem. 7, 87–126.

Nickerson, A., and Raper, K. B. (1973). Macrocysts in the life cycle of the Dictyosteliaceae. I. Formation of the macrocysts. Amer. J. Bot. 60, 190–197.

Niklowitz, W. (1957). Über den Feinbau der Mitochondrien des Schleimpilzes Badhamia utricularis. Exp. Cell Res. 13, 591–595.

Olive, E. W. (1901). A preliminary enumeration of the Sorophoreae. Proc. Amer. Acad. Arts Sci. 37, 333–344.

Olive, E. W. (1902). Monograph of the Acrasieae. Proc. Boston Soc. Natur. Hist. 30, 451–513.

Olive, E. W. (1907). Cytological studies on Ceratiomyxa. Trans. Wis. Acad. Sci., Arts. Lett. 15, 753–774.

Olive, L. S. (1960). Echinostelium minutum. Mycologia 52, 159–161.

Olive, L. S. (1963). The question of sexuality in cellular slime molds. Bull. Torrey Bot. Club 90, 144–147.

Olive, L. S. (1964a). A new member of the Mycetozoa. Mycologia 56, 885–896.

Olive, L. S. (1964b). Spore discharge mechanism in basidiomycetes. *Science* **146**, 542–543.

Olive, L. S. (1965). A developmental study of *Guttulinopsis vulgaris* (Acrasiales). *Amer. J. Bot.* **52**, 513–519.

Olive, L. S. (1967). The Protostelida—a new order of the Mycetozoa. *Mycologia* **59**, 1–29.

Olive, L. S. (1969). Reassignment of Gymnomycota. *Science* **164**, 857.

Olive, L. S. (1970). The Mycetozoa: A revised classification. *Bot. Rev.* **36**, 59–87.

Olive, L. S., and Stoianovitch, C. (1960). Two new members of the Acrasiales. *Bull. Torrey Bot. Club* **87**, 1–20.

Olive, L. S., and Stoianovitch, C. (1966a). A simple new mycetozoan with ballistospores. *Amer. J. Bot.* **53**, 344–349.

Olive, L. S., and Stoianovitch, C. (1966b). *Schizoplasmodium*, a mycetozoan genus intermediate between *Cavostelium* and *Protostelium*; a new order of Mycetozoa. *J. Protozool.* **13**, 164–171.

Olive, L. S., and Stoianovitch, C. (1966c). A new two-spored species of *Cavostelium* (Protostelida). *Mycologia* **58**, 440–451.

Olive, L. S., and Stoianovitch, C. (1966d). *Protosteliopsis*, a new genus of the Protostelida. *Mycologia* **58**, 452–455.

Olive, L. S., and Stoianovitch, C. (1969). Monograph of the genus *Protostelium*. *Amer. J. Bot.* **56**, 979–988.

Olive, L. S., and Stoianovitch, C. (1971a). A new genus of protostelids showing affinities with *Ceratiomyxa*. *Amer. J. Bot.* **58**, 32–40.

Olive, L. S., and Stoianovitch, C. (1971b). *Planoprotostelium*, a new genus of protostelids. *J. Elisha Mitchell Sci. Soc.* **87**, 115–119.

Olive, L. S., and Stoianovitch, C. (1971c). A minute new *Echinostelium* with protostelid affinities. *Mycologia* **63**, 1051–1062.

Olive, L. S., and Stoianovitch, C. (1972). *Protosporangium:* A new genus of protostelids. *J. Protozool.* **19**, 563–571.

Olive, L. S., and Stoianovitch, C. (1974). A cellular slime mold with flagellate cells. *Mycologia* **66**, 685–690.

Olive, L. S., Dutta, S. K., and Stoianovitch, C. (1961). Variation in the cellular slime mold *Acrasis rosea*. *J. Protozool.* **8**, 467–472.

Oosawa, F., Kasae, M., Hatano, S., and Asakura, S. (1966). Polymerization of actin and flagellin. *In* "Principles of Biomolecular Organization" (G. E. W. Wolstenholme and M. O'Connor, eds.), pp. 273–307. Little, Brown, Boston, Massachusetts.

Osborne, T. G. B. (1911). *Spongospora subterranea* (Wallroth) Johnson. *Ann. Bot. (London)* **25**, 327–341.

Park, D., and Robinson, P. M. (1967). Internal water distribution and cytoplasmic streaming in *Physarum polycephalum*. *Ann. Bot. (London)* **31**, 731–738.

Perkins, F. O. (1968). Fine structure of zoospores from *Labyrinthomyxa* sp., parasitizing the clam *Macoma balthica*. *Chesapeake Sci.* **9**, 198–208.

Perkins, F. O. (1969). Ultrastructure of vegetative stages in *Labyrinthomyxa marina* (= *Dermocystidium marinum*), a commercially significant oyster pathogen. *J. Invertebr. Pathol.* **13**, 199–222.

Perkins, F. O. (1970). Formation of centrioles and centriole-like structures during meiosis and mitosis in *Labyrinthula* sp. (Rhizopodea, Labyrinthulida). *J. Cell Sci.* **6**, 629–653.

Perkins, F. O. (1972). The ultrastructure of holdfasts, "rhizoids," and "slime tracks" in thraustochytriaceous fungi and *Labyrinthula* spp. *Arch. Mikrobiol.* **84**, 95–118.

Perkins, F. O. (1973a). A new species of marine labyrinthulid *Labyrinthuloides yorkensis* gen. nov., spec. nov., cytology and fine structure. *Arch. Mikrobiologie* **90**, 1–17.

Perkins, F. O. (1973b). Observations of thraustochytriaceous (Phycomycetes) and labyrinthulid (Rhizopodea) ectoplasmic nets on natural and artificial substrates—an electron microscope study. *Can. J. Bot.* **51**, 485–491.

Perkins, F. O. (1974a). Fine structure of lower marine and estuarine fungi. *In* "Recent Advances in Aquatic Mycology" (E. B. G. Jones, ed.). Paul Elek, London (in press).

Perkins, F. O. (1974b). Phylogenetic considerations of the problematic thraustochytriaceous-labyrinthulid-*Dermocystidium* complex based on observations of fine structure. *Veroeff. Inst. Meeresforsch. Bremerhaven* (in press).

Perkins, F. O. (1974c). Reassignment of *Labyrinthula minuta* to the genus *Labyrinthuloides, Mycologia* **66**, 697–702.

Perkins, F. O., and Amon, J. P. (1969). Zoosporulation in *Labyrinthula sp.*, an electron microscope study. *J. Protozool.* **16**, 235–256.

Perkins, F. O., and Menzel, R. W. (1966). Morphological and cultural studies of a motile stage in the life cycle of *Dermocystidium marinum. Proc. Nat. Shellfish. Ass.* **56**, 23–30.

Perkins, F. O., and Menzel, R. W. (1967). Ultrastructure of sporulation in the oyster pathogen *Dermocystidium marinum. J. Invertebr. Pathol.* **9**, 205–229.

Poche, F. (1913). Das System der Protozoa. *Arch. Protistenk.* **30**, 125–310.

Poff, K. L., and Loomis, W. F., Jr. (1973). Control of phototactic migration in *Dictyostelium discoideum. Exp. Cell Res.* **82**, 236–240.

Pokorny, K. S. (1967). *Labyrinthula. J. Protozool.* **14**, 697–708.

Pokorny, K. S. (1971). The marine protist *Labyrinthula* Cienkowski: A cytochemical and fine structural study of its morphology and secretion of extracellular material. Ph.D. Thesis, Columbia University, New York.

Porter, D. (1969). Ultrastructure of *Labyrinthula. Protoplasma* **67**, 1–19.

Porter, D. (1972). Cell division in the marine slime mold, *Labyrinthula* sp., and the role of the bothrosome in extracellular membrane production. *Protoplasma* **74**, 427–448.

Porter, D. (1974). Phylogenetic considerations of the Thraustochytriaceae and Labyrinthulaceae. *Veroeff. Inst. Meeresforsch. Bremerhaven* (in press).

Potts, G. (1902). Zur Physiologie des *Dictyostelium mucoroides. Flora (Jena)* **91**, 281–347.

Poulos, A., LeStourgeon, W. M., and Thompson, G. A., Jr. (1971). Ether-containing lipids of the slime mold, *Physarum polycephalum*. 1. Characterization and quantification. *Lipids* **6**, 466–469.

Poulter, R. T. M., and Dee, J. (1968). Segregation of factors controlling fusion between plasmodia of the true slime mould, *Physarum polycephalum. Genet. Res.* **12**, 71–79.

Quick, J. A., Jr. (1974). A new marine *Labyrinthula* with unusual locomotion. *Trans. Amer. Microsc. Soc.* **93**, 52–70.

Rai, J. N., and Tewari, J. P. (1963). Studies in cellular slime moulds from Indian soils. II. *Proc. Indian Acad. Sci.* **58**, 201–206.

Rakoczy, L. (1973). The myxomycete *Physarum nudum* as a model organism for photobiological studies. *Ber. Deut. Bot. Ges.* **86**, 141–164 and 178–179.

Rao, A. S., and Brakke, M. K. (1969). Relation of soil-borne wheat mosaic virus and its fungal vector *Polymyxa graminis. Phytopathology* **59,** 581–587.

Raper, J. R. (1966). "Genetics of Sexuality in Higher Fungi." Ronald Press, New York.

Raper, K. B. (1935). *Dictyostelium discoideum,* a new species of slime mold from decaying forest leaves. *J. Agr. Res.* **50,** 135–147.

Raper, K. B. (1940). Pseudoplasmodium formation and organization in *Dictyostelium discoideum. J. Elisha Mitchell Sci. Soc.* **56,** 241–282.

Raper, K. B. (1941a). Developmental patterns in simple slime molds. Third Growth Symposium. *Growth* **5,** 41–76.

Raper, K. B. (1941b). *Dictyostelium minutum,* a second new species of slime mold from decaying forest leaves. *Mycologia* **33,** 633–649.

Raper, K. B. (1951). Isolation, cultivation, and conservation of simple slime molds. *Quart. Rev. Biol.* **26,** 169–190.

Raper, K. B. (1956). *Dictyostelium polycephalum:* A new cellular slime mould with coremiform fructifications. *J. Gen. Microbiol.* **14,** 716–732.

Raper, K. B. (1960). Levels of cellular interaction in amoeboid populations. *Proc. Amer. Phil. Soc.* **104,** 579–604.

Raper, K. B. (1973). Class Acrasiomycetes. *In* "The Fungi" (G. C. Ainsworth, F. K. Sparrow, and A. S. Sussman, eds.), Vol. 4B, pp. 9–36. Academic Press, New York.

Raper, K. B., and Alexopoulos, C. J. (1973). A myxomycete with a singular myxamoebal encystment stage. *Mycologia* **65,** 1284–1295.

Raper, K. B., and Cavender, J. C. (1968). *Dictyostelium rosarium:* A new cellular slime mold with beaded sorocarps. *J. Elisha Mitchell Sci. Soc.* **84,** 31–47.

Raper, K. B., and Fennell, D. (1952). Stalk formation in *Dictyostelium. Bull. Torrey Bot. Club* **79,** 25–51.

Raper, K. B., and Fennell, D. (1967). The crampon-based Dictyostelia. *Amer. J. Bot.* **54,** 315–328.

Raper, K. B., and Quinlan, M. S. (1958). *Acytostelium leptosomum:* A unique cellular slime mould with an acellular stalk. *J. Gen. Microbiol.* **18,** 16–32.

Raper, K. B., and Thom, C. (1941). Interspecific mixtures in the Dictyosteliaceae. *Amer. J. Bot.* **28,** 69–78.

Raven, P. H., and Curtis, H. (1971). "Biology of Plants." Worth, New York.

Reinhardt, D. J. (1966). Silica gel as a preserving agent for the cellular slime mold *Acrasis rosea. J. Protozool.* **13,** 225–226.

Reinhardt, D. J. (1968a). The effects of light on the development of the cellular slime mold *Acrasis rosea. Amer. J. Bot.* **55,** 77–86.

Reinhardt, D. J. (1968b). Development of the mycetozoan *Echinosteliopsis oligospora. J. Protozool.* **15,** 480–493.

Reinhardt, D. J., and Olive, L. S. (1966). *Echinosteliopsis,* a new genus of the Mycetozoa. *Mycologia* **58,** 966–970.

Rhea, R. P. (1966a). Electron microscopic observations on the slime mold *Physarum polycephalum* with specific reference to fibrillar structures. *J. Ultrastruct. Res.* **15,** 349–379.

Rhea, R. P. (1966b). Microcinematographic, electron microscope, and electrophysiological studies on shuttle streaming in the slime mold *Physarum polycephalum. In* "Dynamics of Fluids and Plasmas" (S. I. Pai, ed.), pp. 35–58. Academic Press, New York.

Riedel, V., and Gerisch, G. (1969). Unterschiede im macromolekülbestand zwischen vegetativen und aggregations reifen zellen von *Dictyostelium discoideum* (Acrasina). *Wilhelm Roux' Arch. Entwicklungsmech.* **162**, 268–295.

Riedel, V., Gerisch, G., Müller, E., and Beug, H. (1973). Defective cyclic adenosine-3',5'-phosphate phosphodiesterase regulation in morphogenetic mutants of *Dictyostelium discoideum. J. Mol. Biol.* **74**, 573–585.

Robertson, A., and Cohen, M. H. (1972). Control of developing fields. *Annu. Rev. Biophys. Bioeng.* **1**, 409–464.

Robertson, A., Cohen, M. H., Drage, D. J., Durston, A. J., Rubin, J., and Wonio, D. (1972a). Cellular interactions in slime-mould aggregation. *In* "Cell Interactions" (L. G. Silvestri, ed.), pp. 299–306. North-Holland Publ., Amsterdam.

Robertson, A., Drage, D. J., and Cohen, M. H. (1972b). Control of aggregation in *Dictyostelium discoideum* by an external periodic pulse of cyclic adenosine monophosphate. *Science* **175**, 333–335.

Robison, G. A., Butcher, R. W., and Sutherland, E. W. (1971). "Cyclic AMP." Academic Press, New York.

Rosen, S. D., Kafka, J. A., Simpson, D. L., and Barondes, S. H. (1973). Developmentally regulated, carbohydrate-binding protein in *Dictyostelium discoideum. Proc. Nat. Acad. Sci. U.S.* **70**, 2554–2557.

Ross, I. K. (1957a). Syngamy and plasmodium formation in the Myxogastres. *Amer. J. Bot.* **44**, 843–850.

Ross, I. K. (1957b). Capillitial formation in the Stemonitaceae. *Mycologia* **49**, 808–819.

Ross, I. K. (1960). Studies on diploid strains of *Dictyostelium discoideum. Amer. J. Bot.* **47**, 54–59.

Ross, I. K. (1961). Further studies on meiosis in the myxomycetes. *Amer. J. Bot.* **48**, 244–248.

Ross, I. K. (1964). Pure cultures of some myxomycetes. *Bull. Torrey Bot. Club* **91**, 23–31.

Ross, I. K. (1966). Chromosome numbers in pure and gross cultures of myxomycetes. *Amer. J. Bot.* **53**, 712–718.

Ross, I. K. (1967a). Syngamy and plasmodium formation in the myxomycete *Didymium iridis. Protoplasma* **64**, 104–119.

Ross, I. K. (1967b). Growth and development of the myxomycete *Perichaena vermicularis.* I. Cultivation and vegetative nuclear divisions. *Amer. J. Bot.* **54**, 617–625.

Ross, I. K. (1967c). Growth and development of the myxomycete *Perichaena vermicularis.* II. Chromosome numbers and nuclear cycles. *Amer. J. Bot.* **54**, 1231–1236.

Ross, I. K. (1967d). Abnormal cell behavior in the heterothallic myxomycete *Didymium iridis. Mycologia* **59**, 235–245.

Ross, I. K. (1968). Nuclear membrane behavior during mitosis in normal and heteroploid myxomycetes. *Protoplasma* **66**, 173–184.

Ross, I. K. (1973). The Stemonitomycetidae, a new subclass of myxomycetes. *Mycologia* **65**, 477–485.

Ross, I. K., and Cummings, R. J. (1967). Formation of amoeboid cells from the plasmodium of a myxomycete. *Mycologia* **59**, 725–732.

Ross, I. K., and Cummings, R. J. (1970). An unusual pattern of multiple cell and nuclear fusions in the heterothallic slime mold *Didymium iridis. Protoplasma* **70**, 281–294.

Ross, I. K., and Sunshine, L. D. (1965). The effect of quinic acid and similar compounds on the growth and development of *Physarum flavicomum* in pure culture. *Mycologia* 57, 360–367.

Ross, I. K., Shipley, G. L., and Cummings, R. J. (1973). Sexual and somatic cell fusions in the heterothallic slime mould *Didymium iridis*. 1. Fusion assay, fusion kinetics and cultural parameters. *Microbios* 7, 149–164.

Rostafinski, J. (1875). "Śluzowce (Mycetozoa)." Paris (with suppl., 1876).

Runyon, E. H. (1942). Aggregation of separate cells of *Dictyostelium* to form a multicellular body. *Collect. Net.* 17, 88.

Rusch, H. P. (1968). Some biochemical events in the life cycle of *Physarum polycephalum*. *Advan. Cell Biol.* 1, 297–327.

Sackin, M. J., and Ashworth, J. M. (1969). An analysis of the distribution of volumes amongst spores of the cellular slime mould *Dictyostelium discoideum*. *J. Gen. Microbiol.* 59, 275–284.

Sakai, Y. (1973). Cell type conversion in isolated prestalk and prespore fragments of the cellular slime mold *Dictyostelium discoideum*. *Develop., Growth & Differentiation* 15, 11–19.

Sakai, A., and Shigenaga, M. (1972). Electron microscopy of dividing cells. IV. Behaviour of spindle microtubules during the nuclear division in the plasmodium of the myxomycete *Physarum polycephalum*. *Chromosoma* 37, 101–116.

Sakai, Y., and Takeuchi, I. (1971). Changes of the prespore specific structure during dedifferentiation and cell type conversion of a slime mold cell. *Develop., Growth & Differentiation* 13, 231–240.

Samuel, E. W. (1961). Orientation and rate of locomotion of individual amebas in the life cycle of the cellular slime mold *Dictyostelium mucoroides*. *Develop. Biol.* 3, 317–335.

Samuels, G. J. (1973). The myxomyceticolous species of *Nectria*. *Mycologia* 65, 401–420.

Sansome, E. R., and Dixon, P. A. (1965). Cytological studies of the myxomycete *Ceratiomyxa fruticulosa*. *Arch. Mikrobiol.* 52, 1–9.

Sansome, E. R., and Sansome, F. W. (1961). Observations on *Ceratiomyxa* in West Africa. *J. West Afr. Sci. Ass.* 7, 93–100.

Sauer, H. W. (1973). Differentiation in *Physarum polycephalum*. In "Microbial Differentiation" (J. M. Ashworth and J. E. Smith, eds.), pp. 375–405. Cambridge Univ. Press, London and New York.

Scheetz, R. W. (1972). The ultrastructure of *Ceratiomyxa fruticulosa*. *Mycologia* 64, 38–54.

Scherffel, A. (1925). Endophytische Phycomyceten-Parasiten der Bacillariaceen und einige neue Monadinen. Ein Beitrag zur Phylogenie der Oomyceten (Schröter). *Arch. Protistenk.* 52, 1–141.

Schiebel, W. (1973). The cell cycle of *Physarum polycephalum*. *Ber. Deut. Bot. Ges.* 86, 11–38 and 169–170.

Schinz, H. (1920). Myxogasteres (Myxomycetes, Mycetozoa). *Rabenhorst's Kryptogamen-Flora* 1 (10), 1–472.

Schmoller, H. (1960). Kultur und Entwicklung von *Labyrinthula coenocystis* n. sp. *Arch. Mikrobiol.* 36, 365–372.

Schmoller, H. (1966). Beitrag zur Kenntnis der Labyrinthulen-Entwicklung. *Arch. Protistenk.* 109, 226–244.

Schmoller, H. (1971). Die Labyrinthulen und ihre Beziehung zu den Amöben. *Naturwissenschaften* 58, 142–146.

Schoknecht, J., and Small, E. B. (1972). Scanning electron microscopy of the acellular slime molds (Mycetozoa-Myxomycetes) and the taxonomic significance of surface morphology of spores. *Trans. Amer. Microsc. Soc.* **91**, 380–410.

Schroeter, J. (1897). Myxomycetes. *Naturl. Pflanzenfam.* **1**(1), 1–141.

Schuster, F. (1964a). Electron microscope observations on spore formation in the true slime mold *Didymium nigripes. J. Protozool.* **11**, 207–216.

Schuster, F. (1964b). Ultrastructural and growth studies on slime molds. *Argonne Nat. Lab.* ANL-6971, 70–74.

Schuster, F. (1965a). Ultrastructure and morphogenesis of solitary stages of true slime molds. *Protistologica* **1**, 49–62.

Schuster, F. (1965b). A deoxyribose nucleic acid component in mitochondria of *Didymium nigripes:* A slime mold. *Exp. Cell Res.* **39**, 329–345.

Schuster, F. (1969). Nuclear degeneration during spore formation in the true slime mold, *Didymium nigripes. J. Ultrastruct. Res.* **29**, 171–181.

Shaffer, B. M. (1956). Acrasin, the chemotactic agent in cellular slime moulds. *J. Exp. Biol.* **33**, 645–657.

Shaffer, B. M. (1961). The cells founding aggregation centers in the slime mould *Polysphondylium violaceum. J. Exp. Biol.* **38**, 833–849.

Shaffer, B. M. (1962). The Acrasina. *Advan. Morphog.* **2**, 109–179.

Simon, H. L., and Henney, H. R., Jr. (1970). Chemical composition of slime from three species of myxomycetes. *FEBS (Fed. Eur. Biochem. Soc.) Lett.* **7**, 80–82.

Singh, B. N. (1947). Studies on soil Acrasieae. 1. Distribution of species of *Dictyostelium* in soils of Great Britain and the effect of bacteria on their development. *J. Gen. Microbiol.* **1**, 11–21.

Sinha, U., and Ashworth, J. M. (1969). Evidence for the existence of elements of a parasexual cycle in the cellular slime mould *Dictyostelium discoideum. Proc. Roy. Soc., Ser. B* **173**, 531–540.

Skupienski, F. X. (1927). Sur le cycle évolutif chez une espèce de myxomycète endosporée. *C. R. Acad. Sci.* **182**, 150–152.

Skupienski, F. X. (1928). Badania bio-cytolicze nad *Didymium difforme.* Czesc pierwsza. (Bio-cytological study of *Didymium difforme.* Part 1.) *Acta Soc. Bot. Pol.* **5**, 255–336.

Slifkin, M., and Bonner, J. T. (1952). The effects of salts and organic solutes on the migration time of the slime mold *Dictyostelium discoideum. Biol. Bull.* **102**, 273–277.

Sorokin, N. (1876). *Bursulla crystallina,* nouveau genre de myxomycètes. *Ann. Sci. Natur., Bot. Biol. Veg.* [6] **3**, 40–45.

Sparrow, F. K. (1936). Biological observations on the marine fungi of Woods Hole waters. *Biol. Bull.* **70**, 236–263.

Sparrow, F. K. (1943). "The Aquatic Phycomycetes, Exclusive of the Saprolegniaceae and *Pythium.*" Univ. of Michigan Press, Ann Arbor.

Sparrow, F. K. (1960). "Aquatic Phycomycetes," 2nd rev. ed. Univ. of Michigan Press, Ann Arbor.

Sparrow, F. K. (1973). Mastigomycotina (zoosporic fungi). *In* "The Fungi" (G. C. Ainsworth, F. K. Sparrow, and A. S. Sussman, eds.), Vol. 4B, pp. 61–73. Academic Press, New York.

Stewart, P. A., and Stewart, B. T. (1959). Protoplasmic movement in slime mold plasmodia: The diffusion drag force hypothesis. *Exp. Cell Res.* **17**, 44–58.

Stewart, P. A., and Stewart, B. T. (1961). Membrane formation during sclerotization of *Physarum polycephalum* plasmodia. *Exp. Cell Res.* **23**, 471–478.

Stey, H. (1968). Nachweis eines bisher unbekannten Organells bei *Labyrinthula*. Z. *Naturforsch. B* **23**, 567.

Stey, H. (1969). Elektronmikroskopische Untersuchung an *Labyrinthula coenocystis* Schmoller. Z. *Zellforsch. Mikrosk. Anat.* **102**, 387–418.

Stiemerling, R. (1971). Die Sklerotisation von *Physarum confertum*. *Cytobiologie* **3**, 127–136.

Straub, J. (1954). Das Licht bei der Auslösung der Fruchtkörperbildung von *Didymium nigripes* und die Überstragung der Lichtwirkung durch Plasma. *Naturwissenschaften* **41**, 219–220.

Sussman, M. (1954). Synergistic and antagonistic interactions between morphogenetically deficient variants of the slime mould *Dictyostelium discoideum*. *J. Gen. Microbiol.* **10**, 110–120.

Sussman, M. (1955). "Fruity" and other mutants of the cellular slime mould, *Dictyostelium discoideum*: A study of developmental aberrations. *J. Gen. Microbiol.* **13**, 295–309.

Sussman, M. (1961a). Cultivation and serial transfer of the slime mold *Dictyostelium discoideum* in liquid nutrient medium. *J. Gen. Microbiol.* **25**, 375–378.

Sussman, M. (1961b). Cellular differentiation in the slime mold. *In* "Growth in Living Systems" (M. X. Zarrow, ed.), pp. 221–239. Basic Books, New York.

Sussman, M. (1966). Some genetic and biochemical aspects of the regulatory program for slime mold development. *Curr. Top. Develop. Biol.* **1**, 61–83.

Sussman, M., and Sussman, R. R. (1962). Ploidal inheritance in *Dictyostelium discoideum*: Stable haploid, stable diploid and metastable strains. *J. Gen. Microbiol.* **28**, 417–429.

Sussman, R. R., and Sussman, M. (1953). Cellular differentiation in Dictyosteliaceae: Heritable modifications of the developmental pattern. *Ann. N.Y. Acad. Sci.* **56**, 949–960.

Sykes, E. E., and Porter, D. (1973). Nutritional studies of *Labyrinthula* sp. *Mycologia* **65**, 1302–1311.

Takeuchi, I. (1960). The correlation of cellular changes with succinic dehydrogenase and cytochrome oxidase activities in the development of the cellular slime molds. *Develop. Biol.* **2**, 343–366.

Takeuchi, I. (1963). Immunochemical and immunohistochemical studies on the development of the cellular slime mold *Dictyostelium mucoroides*. *Develop. Biol.* **8**, 1–26.

Takeuchi, I. (1972). Differentiation and dedifferentiation in cellular slime molds. *Aspects Cell. Mol. Physiol.*, pp. 217–236.

Takeuchi, I., and Sakai, Y. (1971). Dedifferentiation of the disaggregated slug cell of the cellular slime mold *Dictyostelium discoideum*. *Develop., Growth & Differentiation* **13**, 201–210.

Therrien, C. D. (1966). Microspectrophotometric measurement of nuclear deoxyribonucleic acid content in two myxomycetes. *Can. J. Bot.* **44**, 1667–1675.

Tommerup, I., and Ingram, D. S. (1971). The life-cycle of *Plasmodiophora brassicae* Woron. in *Brassica* tissue cultures and in intact roots. *New Phytol.* **70**, 327–332.

Toyama, A., and Kusaba, T. (1970). Transmission of soil-borne barley yellow mosaic virus. 2. *Polymyxa graminis* Led., as vector. *Ann. Phytopathol. Soc. Jap.* **36**, 223–229.

Ts'o, P. O., Eggman, L., and Vinograd, J. (1957). Physical and chemical studies of myxomyosin, an ATP-sensitive protein in cytoplasm. *Biochim. Biophys. Acta* **25**, 532–542.

Usui, N. (1971). Fibrillar differentiation in a microplasmodium of the slime mold *Physarum polycephalum. Develop., Growth & Differentiation* **13**, 241–255.

Valkanov, A. (1929). Protistenstudien. 4. Die Natur und die systematische Stellung der Labyrinthuleen. *Arch. Protistenk.* **67**, 110–121.

Valkanov, A. (1969). *Labyrinthodictyon magnificum,* a new rhizoplasmodial unicellular organism. *Progr. Protozool., Abstr. 3rd Int. Congr. Protozool. (Leningrad), 1969* pp. 373–374. Nauka, USSR.

Valkanov, A. (1972). Untersuchungen über die Struktur und den Entwicklungszyklus von *Labyrinthodictyon magnificum* Valk. *Arch. Protistenk.* **114**, 426–443.

van Tieghem, M. P. (1880). Sur quelques myxomycètes à plasmode agrégé. *Bull. Soc. Bot. Fr.* [2] **27**, 317–322.

van Tieghem, M. P. (1884). *Coenonia,* genre nouveau de myxomycètes a plasmode agrégé. *Bull. Soc. Bot. Fr.* [2] **31**, 303–306.

Vishniac, S. H. (1955a). The nutritional requirements of isolates of *Labyrinthula* sp. *J. Gen. Microbiol.* **12**, 455–463.

Vishniac, S. H. (1955b). The activity of steroids as growth factors for a *Labyrinthula* sp. *J. Gen. Microbiol.* **12**, 464–472.

Vishniac, S. H., and Watson, S. W. (1953). The steroid requirements of *Labyrinthula vitellina* var. *pacifica. J. Gen. Microbiol.* **8**, 248–255.

Vogel, H. J. (1964). Distribution of lysine pathways among fungi: Evolutionary implications. *Amer. Natur.* **48**, 435–446.

von Stosch, H. A. (1935). Untersuchungen über die Entwicklungsgeschichte der Myxomyceten. Sexualität und Apogamie bei Didymiaceen. *Planta* **23**, 623–656.

von Stosch, H. A. (1937). Über den Generationswechsel der Myxomyceten; eine Erwiderung. *Ber. Deut. Bot. Ges.* **55**, 362–369.

Ward, J. M., and Havir, E. A. (1957). The role of 3:4-dihydroxytoluene, sulfhydryl groups, and cresolase during melanin formation in a slime mold. *Biochim. Biophys. Acta* **25**, 440–442.

Waterhouse, G. (1973). Plasmodiophoromycetes. *In* "The Fungi" (G. C. Ainsworth, F. K. Sparrow, and A. S. Sussman, eds.), Vol. 4B, pp. 75–82. Academic Press, New York.

Watson, S. W. (1957). Cultural and cytological studies on species of *Labryrinthula.* Ph.D. Thesis, University of Wisconsin, Madison.

Watson, S. W., and Ordal, E. J. (1951). Studies on *Labyrinthula. Univ. Wash. Oceanogr. Lab., Tech. Rep.* **3**, 1–37.

Watson, S. W., and Ordal, E. J. (1957). Techniques for the isolation of *Labyrinthula* and *Thraustochytrium* in pure culture. *J. Bacteriol.* **73**, 589–590.

Watson, S. W., and Raper, K. B. (1957). *Labyrinthula minuta* sp. nov. *J. Gen. Microbiol.* **17**, 368–377.

Watts, D. J., and Ashworth, J. M. (1970). Growth of myxamoebae of the cellular slime mould *Dictyostelium discoideum* in axenic culture. *Biochem. J.* **119**, 171–174.

Weber, A. T., and Raper, K. B. (1972). Induction of fruiting in two aggregateless mutants of *Dictyostelium discoideum. Develop. Biol.* **26**, 606–615.

Weinkauff, A. M., and Filosa, M. F. (1965). Factors involved in the formation of macrocysts by the cellular slime mold, *Dictyostelium mucoroides. Can. J. Microbiol.* **11**, 385–387.

Weitzman, I. (1962). Studies on the nutrition of *Acrasis rosea. Mycologia* **54**, 113–115.

Welden, A. (1955). Capillitial development in the myxomycetes *Badhamia gracilis* and *Didymium iridis*. *Mycologia* **47**, 714–728.

Wheals, A. E. (1970). A homothallic strain of *Physarum polycephalum*. *Genetics* **66**, 623–633.

Wheals, A. E. (1973). Developmental mutants in a homothallic strain of *Physarum polycephalum*. *Genet. Res.* **21**, 79–86.

Whiffen, A. (1939). The cytology of a new species of the Plasmodiophoraceae. *J. Elisha Mitchell Sci. Soc.* **55**, 243 (abstr.).

Whittaker, R. H. (1969). New concepts of kingdoms of organisms. *Science* **163**, 150–160.

Williams, K. L., Kessin, R. H., and Newell, P. C. (1974). Genetics of growth in axenic medium of the cellular slime mould *Dictyostelium discoideum*. *Nature (London)* **247**, 142–143.

Williams, P. H., and McNabola, S. (1967). Fine structure of *Plasmodiophora brassicae* in sporogenesis. *Can. J. Bot.* **45**, 1665–1669.

Williams, P. H., and McNabola, S. (1970). Fine structure of the host-parasite interface of *Plasmodiophora brassicae* in cabbage. *Phytopathology* **60**, 1557–1561.

Williams, P. H., and Yukawa, Y. B. (1967). Ultrastructural studies on the host-parasite relations of *Plasmodiophora brassicae*. *Phytopathology* **57**, 682–687.

Williams, P. H., Reddy, M. N., and Strandberg, J. O. (1969). Growth of noninfected and *Plasmodiophora brassicae* infected cabbage callus in culture. *Can. J. Bot.* **47**, 1217–1221.

Wilson, C. M. (1953). Cytological study of the life cycle of *Dictyostelium*. *Amer. J. Bot.* **40**, 714–718.

Wilson, C. M., and Cadman, E. J. (1928). The life history and cytology of *Reticularia lycoperdon*. *Trans. Roy Soc. Edinburgh* **55**, 555–608.

Wilson, C. M., and Ross, I. K. (1955). Meiosis in the myxomycetes. *Amer. J. Bot.* **42**, 743–749.

Wilson, C. M., and Ross, I. K. (1957). Further cytological studies in the Acrasiales. *Amer. J. Bot.* **44**, 345–350.

Wohlfarth-Bottermann, K. E. (1964). Differentiations of the ground cytoplasm and their significance for the generation of the motive force of amoeboid movement. *In* "Primitive Motile Systems in Cell Biology" (R. D. Allen and N. Kamiya, eds.), pp. 79–109. Academic Press, New York.

Wollman, C., and Alexopoulos, C. J. (1967). The plasmodium of *Licea biforis* in agar culture. *Mycologia* **59**, 423–430.

Wright, B. (1973). "Critical Variables in Differentiation," pp. 26–52. Prentice-Hall, Englewood Cliffs, New Jersey.

Yabuno, K. (1971). Changes in cellular adhesiveness during the development of the slime mold *Dictyostelium discoideum*. *Develop., Growth & Differentiation* **13**, 181–190.

Yamada, T., Yanagisawa, K. O., Ono, H., and Yanagisawa, K. (1973). Genetic analysis of developmental stages of the cellular slime mold *Dictyostelium purpureum*. *Proc. Nat. Acad. Sci. U.S.* **70**, 2003–2005.

Yanagisawa, K., Loomis, W. F., Jr., and Sussman, M. (1967). Developmental regulation of the enzyme UDP-galactose polysaccharide transferase. *Exp. Cell Res.* **46**, 328–334.

Yemma, J. J., and Therrien, C. D. (1972). Quantitative microspectrophotometry of nuclear DNA in selfing strains of the myxomycete *Didymium iridis*. *Amer. J. Bot.* **59**, 828–835.

Young, E. L. (1937). Notes on the labyrinthulan parasite of eel-grass *Zostera marina. Bull. Mt. Desert Island Biol. Lab.* pp. 33–35.

Young, E. L. (1943). Studies on *Labyrinthula.* The etiologic agent of the wasting disease of eel-grass. *Amer. J. Bot.* 30, 586–593.

Zopf, W. (1885). Die Pilzthiere oder Schleimpilze. *Encykl. Naturwiss.* 3, 1–174.

Zopf, W. (1892). Zur Kenntnis der Labyrinthuleen, einer Familie der Mycetozoen. *Beitr. Physiol. Morphol. Nied. Organ.* 2, 36–48.

Index to Genera and Species of Mycetozoans and Associates

A

Acrasis, 167, 169, 170, 174, 175, 180, 181, 249, 250
Acrasis granulata, 167, 169, 174, 180
Acrasis rosea, 11, 160–169, 171, 174–176, 179, 184–189, 249, 250
Acytostelium, 11, 51–54, 61, 82, 84–86, 90, 98, 248
Acytostelium ellipticum, 52
Acytostelium irregularosporum, 52
Acytostelium leptosomum, 52–54, 61, 84–86, 248
Althornia, 241
Althornia crouchii, 241
Aplanochytrium, 216, 224, 230
Aplanochytrium kerguelensis, 230
Arcyria, 112, 113
Arcyria cinerea, 112, 113, 126, 141, 145
Arcyria incarnata, 148

B

Badhamia affinis, 120
Badhamia curtisii, 103, 121
Badhamia gracilis, 138
Badhamia utricularis, 132, 140
Barbeyella, 110
Bursulla crystallina, 111, 112

C

Cavostelium, 19, 20, 29, 32, 38, 43, 150, 246
Cavostelium apophysatum, 20, 21, 32
Cavostelium bisporum, 16, 22, 38, 39, 41, 43, 110, 135, 151, 246
Ceratiomyxa, 12, 19, 21, 27, 29–32, 43
Ceratiomyxa fruticulosa, 17, 29–32, 43, 150
Ceratiomyxa morchella, 32
Ceratiomyxa sphaerosperma, 32
Ceratiomyxella, 19, 20, 21, 35, 38, 246
Ceratiomyxella tahitiensis, 21, 23, 24, 25, 34, 38, 41–43
Ceratiomyxella tahitiensis var. *neotropicalis,* 21
Chlamydomyxa, 215
Clastoderma, 110
Clastoderma debaryanum, 144
Coenonia, 51, 67, 248
Coenonia denticulata, 67
Comatricha, 114, 136, 138
Comatricha typhoides, 114
Copromyxa, 98, 170, 176–180, 181, 184, 189, 190, 249, 250
Copromyxa arborescens, 160, 162, 177–180, 187–190, 249, 250
Copromyxa protea, 169, 177
Cribraria, 118
Cribraria sp., 119

D

Dermocystidium marinum, 227
Diachea, 113
Diachea leucopodia, 113, 114

Dictydiaethalium, 119, 247
Dictydiaethalium plumbeum, 119
Dictydium, 118
Dictydium cancellatum, 118–120
Dictyostelium, 1, 51, 52, 54–65, 67, 83, 85, 86, 169, 248
Dictyostelium aureum, 56
Dictyostelium brevicaule, 56
Dictyostelium coeruleo-stipes, 56, 63
Dictyostelium delicatum, 63
Dictyostelium deminutivum, 55, 61
Dictyostelium dimigraformum, 55, 56
Dictyostelium discoideum, 1, 47–51, 55–60, 67–84, 88–92, 95–98, 125, 190
Dictyostelium firmibasis, 63
Dictyostelium giganteum, 55, 59
Dictyostelium irregularis, 49, 55
Dictyostelium lacteum, 55, 61
Dictyostelium laterosorum, 55, 56, 62
Dictyostelium lavandulum, 56, 62
Dictyostelium minutum, 55, 60, 71, 74, 84, 92
Dictyostelium monochasioides, 64
Dictyostelium mucoroides, 45, 46, 48, 55–60, 64, 65, 68, 74–76, 78, 82, 84, 92, 94, 97
Dictyostelium mucoroides var. *stolonifera*, 59
Dictyostelium polycephalum, 55, 59, 61, 62, 65, 74, 78
Dictyostelium purpureum, 49, 55, 57, 60, 68, 74–78, 84, 92, 95, 97
Dictyostelium rhizopodium, 56, 62, 64
Dictyostelium rosarium, 55, 62, 63, 74
Dictyostelium roseum, 56
Dictyostelium sphaerocephalum, 56, 63
Dictyostelium vinaceo-fuscum, 56, 63
Diderma, 117
Diderma floriforme, 117
Didymium, 117, 138
Didymium difforme, 152
Didymium iridis, 117, 121–124, 133, 136, 142, 150, 152–156
Didymium nigripes, 6, 103, 121–123, 134, 136, 137, 140, 143, 146, 147
Didymium squamulosum, 123, 144
Diplophrys stercorea, 227, 228

E

Echinosteliopsis oligospora, 110, 111
Echinostelium, 20, 29, 43, 109–111, 135, 138, 246, 247
Echinostelium lunatum, 29, 109, 110, 115, 121, 135, 150, 151, 246, 247
Echinostelium minutum, 109, 110, 120, 123, 140, 141, 147, 150, 151
Endemosarca, 201, 203, 206, 212–214
Endemosarca anomala, 203, 206
Endemosarca hypsalyxis, 203, 204, 205, 212–214
Endemosarca ubatubensis, 203, 204, 205, 212, 213

F

Fuligo, 116, 137
Fuligo septica, 116, 117, 144, 152

G

Guttulina, 169–171, 173, 177, 180, 181
Guttulina aurea, 173
Guttulina protea, 169, 176, 177
Guttulina rosea, 169
Guttulina sessilis, 173
Guttulina sp., 184
Guttulinopsis, 169, 170, 174, 180–184, 249
Guttulinopsis clavata, 184
Guttulinopsis stipitata, 184
Guttulinopsis vulgaris, 160, 162, 170, 180–184, 189, 190, 227, 249

H

Hemitrichia, 138, 247
Hemitrichia clavata, 120
Hemitrichia serpula, 112, 113

J

Japonochytrium, 216, 224, 230, 241
Japonochytrium marinum, 230, 241
Japonochytrium sp., 238

L

Labyrinthodictyon, 215
Labyrinthodictyon magnificum, 226
Labyrinthomyxa, 215, 227
Labyrinthomyxa marina, 227
Labyrinthomyxa pohlia, 227
Labyrinthomyxa sauvageaui, 227
Labyrinthorhiza, 215
Labyrinthula, 2, 5, 215, 217, 218, 220–239, 241, 251
Labyrinthula algeriensis, 218, 222, 223, 225
Labyrinthula chattonii, 226
Labyrinthula cienkowskii, 226
Labyrinthula coenocystis, 225, 234
Labyrinthula macrocystis, 215, 222, 226
Labyrinthula magnifica, 227
Labyrinthula roscoffensis, 226
Labyrinthula saliens, 225
Labyrinthula sp., 219–221, 232, 233, 236–238
Labyrinthula thaisi, 225
Labyrinthula valkanovii, 226
Labyrinthula vitellina, 215, 217, 218, 225, 236, 237
Labyrinthula vitellina var. *pacifica,* 223, 225
Labyrinthula zopfi, 226
Labyrinthuloides, 216, 221, 224, 228, 229, 231, 241, 251
Labyrinthuloides minuta, 223, 229, 238
Labyrinthuloides yorkensis, 228, 229, 238
Lamproderma, 114, 136
Lamproderma arcyrionema, 114
Leocarpus, 116
Leocarpus fragilis, 116, 117
Licea, 118, 135
Licea biforis, 118
Licea sp., 118
Ligniera, 194, 202
Ligniera betae, 194
Ligniera junci, 194, 199, 200
Lycogala, 119
Lycogala epidendrum, 119
Lycogala flavofuscum, 120

M

Metatrichia vesparium, 112, 113

N

Nematostelium, 19, 25, 32, 34, 35
Nematostelium gracile, 34
Nematostelium ovatum, 17, 34, 35

O

Octomyxa, 202
Octomyxa achlyae, 199

P

Perichaena, 112
Perichaena depressa, 112
Perichaena vermicularis, 122, 125, 150
Phagomyxa, 203, 206
Phagomyxa algarum, 201, 203
Physarella oblonga, 103
Physarum, 48, 115, 126, 140, 147
Physarum cinereum, 103, 142, 143
Physarum compressum, 103, 126
Physarum flavicomum, 103, 141–143, 147, 149, 152, 153
Physarum gyrosum, 122, 125
Physarum nudum, 134
Physarum polycephalum, 1, 102, 103, 115–117, 123–141, 144, 152, 154–156, 247
Physarum pusillum, 123, 141, 148, 152
Physarum rigidum, 103, 152, 153
Planoprotostelium, 19, 20, 25
Planoprotostelium aurantium, 18, 19, 25, 36
Plasmodiophora, 2, 202
Plasmodiophora brassicae, 194, 195, 197–200, 206, 208, 211, 212
Pocheina, 169, 170, 171–174, 180, 181, 250
Pocheina rosea, 160, 161, 169, 171–173, 181, 185, 249, 250
Polymyxa, 202
Polymyxa betae, 196, 208, 211
Polymyxa graminis, 200

Polysphondylium, 51, 57, 59, 62, 64–67, 74, 84, 86, 94, 95, 248
Polysphondylium album, 65
Polysphondylium candidum, 65
Polysphondylium pallidum, 46, 47, 65, 66, 74, 97, 98
Polysphondylium violaceum, 50, 65, 71, 74, 75, 78, 84, 86, 87, 92–95
Protosporangium, 19, 20, 25, 27–29, 246
Protosporangium articulatum, 27, 28
Protosporangium bisporum, 26–28, 32
Protosporangium fragile, 27, 28
Protosteliopsis, 19, 38
Protosteliopsis fimicola, 37, 38, 40, 41
Protostelium, 11, 19, 32, 35–38, 248
Protostelium arachisporum, 36, 37
Protostelium irregularis, 14, 37, 248
Protostelium mycophaga, 11, 13, 14, 18, 25, 35, 36
Protostelium pyriformis, 37
Protostelium zonatum, 37
Pseudoplasmodium, 215

R

Reticularia lycoperdon, 120, 140

S

Schizochytrium, 216, 221, 224, 230, 239, 241, 251
Schizochytrium aggregatum, 230, 238, 241
Schizoplasmodiopsis, 19, 33
Schizoplasmodiopsis pseudoendospora, 34
Schizoplasmodium, 19, 32, 34
Schizoplasmodium cavostelioides, 17, 32, 33
Sorodiscus, 202
Sorosphaera, 202

Sorosphaera veronicae, 195, 197–200, 207–210
Spongospora, 202
Spongospora subterranea, 194, 195, 199
Stemonitis, 113, 114, 136, 138, 140, 145, 247
Stemonitis axifera, 114
Stemonitis fusca, 144
Stemonitis herbatica, 126, 136, 147
Stemonitis sp., 116, 132
Stemonitis virginiensis, 138, 139, 146, 147, 151–153

T

Tetramyxa, 202
Thraustochytrium, 216, 221, 224, 229, 230, 239, 241, 251
Thraustochytrium aggregatum, 229
Thraustochytrium amoebidum, 229
Thraustochytrium antarcticum, 229
Thraustochytrium aureum, 229, 230, 239
Thraustochytrium globosum, 229
Thraustochytrium kerguelensis, 229
Thraustochytrium kinnei, 229
Thraustochytrium motivum, 222, 229, 238–241
Thraustochytrium multirudimentale, 229
Thraustochytrium pachydermum, 229
Thraustochytrium proliferum, 229
Thraustochytrium roseum, 229, 230
Thraustochytrium rossii, 229
Thraustochytrium sp., 240, 241
Thraustochytrium striatum, 229
Thraustochytrium visurgensis, 229
Trichia, 113, 138, 247
Trichia sp., 116
Trichia varia, 112, 113

W

Woronina, 202

Subject Index

A

Acrasea, 159–190, 249, 250, 252
  in key, 5
Acrasia, 2, 170
Acrasida, 159, 170
  key to families and genera of, 170
Acrasidae, 170–176
  in key, 170
Acrasids, 159–190, 248–250
  classification of, 169–170
  distribution of, 161, 162
  growth variants of, 175
  isolation and culture of, 161, 162, 174, 175
  life cycle of, 162–169
  phylogeny of, 248–250
  ultrastructure of, 184–190
Acrasieae, 2, 45, 159, 169
Acrasin, 50, 72–74. *See also* Cyclic AMP
Acrasiomycetes, 5
Actomyosin, *see* Proteins of myxomycetes, contractile
Acytosteliidae, 52–54
  in key, 51
Adhesiveness of cells in dictyostelids, 71, 72, 79, 89, 90
Aggregation
  in acrasids, 160, 164, 168, 179, 183
  in dictyostelids, 50, 53, 61, 64, 68–76, 86, 88
    antigens influencing, 71
    distribution of centers of, 75, 76
    varied responses to cyclic AMP, 74

of spindle cells in *Labyrinthula*, 222, 226
Amoebae
  of acrasids, 159, 160, 163, 164, 171–173, 176, 179, 181
  of dictyostelids, 47–50, 67–76
    anastomosis of, 72
    plurinucleate, 72
  of myxomycetes, 121–123
    nonsexual cell fusions among, 123
  of protostelids, 11, 14, 15, 35–38, 40, 41
  types of, 6, 248, 249
Antigens, cell surface, of *Dictyostelium*, 71
Aphelidiopsis, 202
Aphelidiopsis epithemiae, 202
Axenic culture
  of dictyostelids, 47
  of *Labyrinthula*, 217, 218
  of myxomycetes, 102, 103
  of thraustochytrids, 218

B

Ballistospores of *Schizoplasmodium*, 17, 33
Bothrosomes, *see* Sagenogens

C

Capillitium of myxomycetes, 105, 107, 116, 138, 139
  ontogeny of, 138
Carotenoids of *Acrasis rosea*, 169

Cavosteliidae, 19–29
  in key, 19
Cellular slime molds, see Acrasids and
  Dictyostelids
Cellulose, evidence of
  in dictyostelids, 54, 76, 91, 92
  in myxomycetes, 147
  in protostelids, 19, 25
Centrioles
  of Endemosarca, 213
  of myxomycetes, 103, 141–143, 147
  of plasmodiophorids, 210, 211
  of protostelids, 41–43
Ceratiomyxidae, 29–32
  in key, 19
Chromosome numbers
  in Ceratiomyxa fruticulosa ($n = 8$), 31
  in Dictyostelium discoideum ($n = 7$),
    48, 49, 70, 96
  in Endemosarca ($n = 5$–6), 203
  in Labyrinthula sp. ($n = 9$), 237
  in myxomycetes, 121, 125
  in Phagomyxa algarum ($n = 5$–6), 203
  in plasmodiophorids, 197
  in Polysphondylium violaceum
    ($n = 8$–9), 48, 50
Ciliates
  parasites of, 203–206
  sorogenesis in, 6, 7
Clastodermidae, 110
Coaggregation in dictyostelids, 71, 76,
    97, 98
Compatibility system of myxomycetes,
    152, 153
Copromyxidae, 170, 176–180
  in key, 170
Cribrariidae, 118
Cruciform division, see Mitosis in
  plasmodiophorids
Culmination, see Sorogenesis
Cyclic AMP
  and Acrasis rosea, 160
  and aggregation in dictyostelids, 50,
    73–76, 86
  in myxomycetes, 130, 131
  in pseudoplasmodium of Dictyoste-
    lium, 79
  and sorogenesis in Dictyostelium, 83
Cystogenesis, see Sporogenesis in
  plasmodiophorids

Cystosori of plasmodiophorids, 199,
  200, 202

D

Dedifferentiation in Dictyostelium, 79
Dianemidae, 112
Dictyostelia, 45–98, 248
  in key, 5
Dictyosteliaceae, 45
Dictyostelids, 45–98, 248
  descriptions of taxa, 52–67
  distribution of, 46, 47
  genetics of, 95–98
  isolation and culture of, 46, 47
  key to families and genera, 51
  life cycle of, 47–51, 67–84
  phylogeny of, 248
  ultrastructure of, 84–94
Dictyosteliida, 45, 51
Dictyosteliidae, 54–67
  in key, 51
Didymiidae, 116
DNA
  base ratios
    in dictyostelids, 59, 74, 248
    in Protostelium, 248
  in cell cycle
    of Dictyostelium, 68
    of Physarum, 126, 127
  hybridization with RNA in Dictyoste-
    lium, 98
  in mitochondria of myxomycetes, 144
  in nuclei of selfed Didymium clones,
    153

E

Echinosteliida, 109–112, 136
  in key, 108
Echinosteliidae, 109, 110
Echinosteliopsida, 110
Echinosteliopsidae, 110, 111
Ectoplasmic net, 216, 218–223, 228–231,
    233–235, 238–240
Encystment of trophic phase
  in acrasids, 164, 177, 178
  in dictyostelids, 92–94
  in Labyrinthula, 222, 226

in myxomycetes, 104, 105, 121, 122, 130–133, 145
in protostelids, 15, 41, 43
Endemosarca, ultrastructure of, 212–214
Endemosarcidae, 201, 203–206
Eumycetozoa, 10–156, 245–249, 252
  in key, 4
Eumycota, 5, 6
Euprotista, 4, 201, 250

F

Flagellate cells
  of an acrasid (*Pocheina rosea*), 171–173
  of *Endemosarca*, 203–205
  of *Labyrinthula*, 223, 224, 236, 237
    ontogeny of, 223, 237
  of myxomycetes, 99, 103, 109, 120–122, 141–143
  of plasmodiophorids, 193, 195–198, 200, 211, 212
    ontogeny of, 196
    as virus vectors, 200
  of proteomyxans, 202, 203
  of protostelids, 15, 16, 18, 20, 21, 23–29, 32, 38, 39, 41–43
  of thraustochytrids, 223, 224, 228–230, 241
Founder cells of *Dictyostelium*, 71
Fruiting, *see* Sorogenesis *and* Sporogenesis
Fungi, 3–6, 201, 215, 223, 247, 249–251
  keys to major taxa of, 3, 4
  lysine pathways in, 4

G

Genetics
  of dictyostelids, 95–98
  of myxomycetes, 152–156
Guttulinaceae, 159, 169
Guttulinopsidae, 170, 180–184
  in key, 170
Gymnococcaceae, 203
Gymnococcidae, 206, 250
Gymnococcus, 202
Gymnococcus cladophorae, 203
Gymnomycota, 3, 5

Gymnomyxa, 2, 4, 5, 245, 252
  key to major taxa of, 4, 5
  phylogeny of, 245–252

H

Heimerliaceae, 206

I

Incompatibility system of myxomycetes, 153–155
Initiator (I) cells, 68, 70, 71

K

Key
  to *Acytostelium*, 52
  to *Dictyostelium*, 55, 56
  to families and genera
    of Acrasida, 170
    of Dictyosteliida, 51
    of Labyrinthulida, 224
    of Protosteliida, 19
  to Fungi, major taxa of, 3, 4
  to Gymnomyxa, major taxa of, 4, 5
  to Mycetozoa, major taxa of, 4, 5
  to Myxogastria, orders of, 108
  to Plasmodiophoridae, genera of, 202
Kinetosomes
  of Labyrinthulina, 237, 241
  of myxomycetes, 141, 142
  of protostelids, 39, 41, 43

L

Labyrinthulas, 215, 225–228
  cell movement in, 221
  distribution of, 216, 217
  isolation and culture of, 216–218
  life cycle of, 218–223
  mitotic cell division in, 222, 235, 237
  pathogenicity of, 217
  phylogeny of, 250–252
  plasmodium-like protoplasts of, 225
  ultrastructure of, 230–237
Labyrinthulales, 5, 6
Labyrinthulea, 5, 224
Labyrinthulia, 2, 225
Labyrinthulida, 224
  key to families and genera of, 224

Labyrinthulidae, 215, 225–227
    doubtful genera of, 227, 228
    in key, 224
Labyrinthulina, 5, 215–241, 250–252
    cell wall structure of, 231, 238, 240,
        241, 251
    in key, 5
    sporangial cytokinesis in, 224
    flagellate cells of, 224
Labyrinthulomycota, 223
Liceida, 117–119
    in key, 108
Liceidae, 118
Life cycle
    of acrasids (Acrasis rosea), 162–169
    of dictyostelids, 47–51, 67–84
    of myxomycetes, 103–105, 120–141
    of plasmodiophorids, 194–200
    of protostelids, 14–18, 21, 23, 26
        Ceratiomyxella tahitiensis, 21, 23
        Protosporangium bisporum, 26
Light, effect of on
    migration in dictyostelids, 78
    sorogenesis in Acrasis rosea, 168, 169,
        176
    sporogenesis in myxomycetes, 133–135

M

Macrocysts of dictyostelids, 84, 92–94
Mastigomycotina, 6, 201
Meiosis
    in Ceratiomyxa, 18, 31
    in dictyostelids, possibility of, 93–95
    in Labyrinthula, 236, 237
    in myxomycetes, 105, 123, 140, 141,
        147–150
    in plasmodiophorids, 195, 198, 199,
        209–211
Microsporangia of Protosporangium,
    25–28
Migration, see Plasmodium and Pseudo-
    plasmodium
Mitochondria characteristics of
    in acrasids, 184–186, 188–190, 249
    in Endemosarca, 213, 214
    in eumycetozoans, 249
    in Labyrinthulina, 231, 232, 238
    in plasmodiophorids, 206, 208–211

Mitochondria and préspore vacuoles, 89
Mitochondrial types
    in fungi, 249
    in mycetozoans, 249
Mitosis
    in acrasids, 164, 180
    in dictyostelids, 48, 76, 86, 87
    in Endemosarca, 203, 213
    in Labyrinthuyla, 222
    in myxomycetes
        during cleavage in Stemonitis, 140,
            145, 146
        in haploid cells, 103, 121, 143
        in plasmodium, 121, 124–127, 143,
            144
        precleavage, 105, 135, 140, 141,
            145, 146
    in Phagomyxa, 203
    in plasmodiophorids, 196–198, 206–208
    in protostelids, 15, 16, 18, 21, 24–26,
        29, 38
    in thraustochytrids, 229
Monadineae, 200
Mutagenesis
    in dictyostelids, 95–97
    in myxomycetes, 156
Mycetozoa, key to major taxa of, 4, 5
Mycetozoia, 2, 201
Myxogastria, 2, 99–156, 246–248. See also
        Myxomycetes
    in key, 5
    key to orders of, 108
Myxomycetes, 2, 3, 5, 45, 99–156,
        246–248
    classification of, 106–108
    description of taxa, 109–119
    distribution of, 100, 101
    genetics of, 152–156
    isolation and culture of, 101–103
    phylogeny of, 246–248
    spore germination in, 120, 141, 151, 152
    ultrastructure of, 141–152
Myxomycota, 5
Myxomycotina, 99

N

Nectria, parasitic on myxomycetes, 101
Neotypes of Acrasis rosea, 175

Net plasmodium, *see* Ectoplasmic net
Nucleus, characteristics of
    in acrasids, 160, 179, 185–190
    in dictyostelids, 47–49, 52, 85, 87, 93, 94
    in *Echinosteliopsis,* 111
    in *Endemosarca,* 213
    in Labyrinthulina, 232
    in myxomycetes, 109, 121, 123, 125, 147, 148
    in protostelids, 14, 38, 43

O

Oomycetes, 3, 4, 238, 250, 251

P

Parasexuality in dictyostelids, 95–97
Phosphodiesterase
    developmental role of, 74–76, 130
    inhibitor of, 75
    membrane-bound, 75
    in mutants of *Dictyostelium,* 75
    in myxomycetes, 130
Phylogeny of the Gymnomyxa, 245–252
Physarida, 115–117
    in key, 108
Physaridae, 115
Phytomyxinae, 200
Pigment granules in *Acrasis,* 184, 186
Pinocytosis, 38, 103
Plasmodiophoraceae, 200
Plasmodiophorales, 201
Plasmodiophorea, 5, 201, 206
Plasmodiophoridae, 201, 202
    key to genera of, 202
Plasmodiophorids, 193–214
    classification of, 200–206
    infection apparatus of, 211, 212
    life cycle of, 194–200
    maintenance of in laboratory, 194
    occurrence of in nature, 194
    phylogeny of, 250
    ultrastructure of, 206–212
Plasmodiophoromycetes, 6, 201
Plasmodiophorina, 5, 193–214, 252
    in key, 5
Plasmodium
    of *Endemosarca,* 203, 205, 212–214

    of myxomycetes, 100, 104–106, 123–133, 143–145
    contractile proteins in, 144, 145
    encystment of, 130–133
    incompatibility system of, 153–155
    microfibrils in, 144
    migration of, 105, 127–130, 136
    ontogeny of, 124–126
    pigments in, 134, 155, 156
    protoplasmic streaming in, 127–130
    types of, 106
    of *Phagomyxa,* 203
    of plasmodiophorids, 193, 195–200, 206
    of proteomyxans, 202
    of protostelids, 23–25, 29, 32–35
Plasmotomy, 15, 16, 50, 72, 126, 180, 197, 229
Preaggregation in dictyostelids, 50, 67, 68, 82
Prespore cells
    of dictyostelids, 49, 50, 78, 80–83, 89–90
    of protostelids, 15, 17, 23, 25, 26, 29–31, 33–35, 43
Prespore vacuoles of dictyostelids, 86, 88, 89
Prestalk cells of dictyostelids, 49, 50, 78, 80–83, 89–91
Promitosis, *see* Mitosis in plasmodiophorids
Proteins of myxomycetes
    contractile, 127–129, 144, 145
    glycoprotein in walls, 131, 147
    mitosis-inducing, 127
    nuclear, 127
Proteomyxa, 2, 193, 200–202, 206, 250
Protista, position of Gymnomyxa in, 4, 5
Protocentrioles in Labyrinthulina, 233, 235, 237
Protoctista, 2
Protoplasta, 2
Protostelia, 2, 6, 11–43, 246, 248
    in key, 5
Protostelids, 11–43, 246, 248
    classification of, 18, 19
    description of taxa, 19–38
    distribution of, 12, 13
    isolation and culture of, 12, 13
    life cycles of, 14–18, 21, 23, 25, 26

phylogeny of, 246, 248
ultrastructure of, 38–43
Protosteliida, 18, 19
key to families and genera of, 19
Protosteliidae, 19, 32–38
in key, 19
Protozoa, 2, 200, 201, 215
Pseudocapillitium of myxomycetes, 107,
   119
Pseudoplasmodium
of acrasids, 161, 164, 165, 180, 181,
   183, 184, 187
of dictyostelids, 49, 50, 53, 61, 62,
   76–79, 88
culmination of, 80–84
migration of, 50, 55, 56, 61, 65, 76–79
prespore and prestalk cells in, 78, 79,
   88
Pseudospores, 170, 181
Pseudosporopsis, 202

R

Reticulariidae, 118
Rhizopoda, 201
Rhizopodea, 2, 201, 215
RNA
hybridization with DNA in *Dictyostelium*, 98
ribosomal, in Labyrinthulina, 224, 251
synthesis in *Physarum*, 127, 131, 135,
   145

S

Sagenogens, 223, 224, 231, 233–235, 237,
   238
Sappinia, 6
Sappiniaceae, 45
Sarcodina, 2, 201, 225
Sarcomastigophora, 2
Sarkodina, 2
Scales
in algal protists, 246, 251, 252
in cell wall of *Labyrinthula*, 231
in cell wall of thraustochytrids, 238,
   239, 241
peripheral, of protostelids, 41–43
peripheral, of thraustochytrid zoospores, 241

Sclerotia of myxomycetes, 130–133, 145
Sexual reproduction
in *Ceratiomyxa*, 17, 18, 32, 43
in dictyostelids, possibility of, 51,
   93–95
in *Labyrinthula*, evidence of, 218, 223,
   237
in myxomycetes, 104, 105, 122, 123,
   152, 153
compatibility system of, 152, 153
in plasmodiophorids, 195–200
Slime sheath
of acrasids, 165, 185
of dictyostelids, 49–51, 76–78, 83, 89,
   90, 92
of myxomycetes, 105, 125, 126, 128,
   135
of protostelids, 16, 17, 30, 33
Slimeways, *see* Ectoplasmic net
Slug, *see* Pseudoplasmodium of dictyostelids
Sorocysts of acrasids, 176–178, 184,
   187–189
Sorogen
of acrasids, 165, 167
of dictyostelids, 49–51, 53, 54, 58, 59,
   61, 64–66, 80–83
Sorogenesis
in acrasids, 162, 164–169, 180, 182–184
   188
Acrasis rosea, 164–169, 188
Copromyxa, 180
Guttulinopsis vulgaris, 182–184
in dictyostelids
Acytostelium leptosomum, 53–54, 85,
   86
Dictyostelium discoideum, 49–51, 57,
   80–84
Dictyostelium mucoroides, 56–59
Dictyostelium polycephalum, 61, 62
Polysphondylium, 57, 64–66
Spacing substance in *Dictyostelium*, 83,
   84
Spindle cells of *Labyrinthula*, 216,
   218–222, 231–235, 237
mitotic division of, 222, 235, 237
cell wall of, 231
Sporangia, *see* Sporocarps of myxomy-
   cetes *and* Zoosporangia
Spore vacuoles of dictyostelids, 89

Spore wall
  composition of
    in dictyostelids, 54, 92
    in myxomycetes, 147, 247, 248
    in plasmodiophorids, 200
  structure of
    in acrasids, 187–190
    in dictyostelids, 54, 92
    in myxomycetes, 147, 150, 151
    in plasmodiophorids, 208
    in protostelids, 39, 43
Sporocarps of myxomycetes, types of, 106, 107
Sporogenesis, *see also* Sorogenesis
  in *Endemosarca*, 203–205, 213, 214
  in *Labyrinthula*, 222, 223
  in myxomycetes, 105, 133–141, 145–150
    factors influencing, 133–135
    spore cleavage, 105, 139, 140, 145–147
    stages of, 136
  in plasmodiophorids, 197–200, 208, 209
  in protostelids, 15–17, 29–31, 37, 43
  in thraustochytrids, 228–230
Stalk development
  in acrasids, 165, 167
  in *Acytostelium*, 53, 54, 84–86
  in *Dictyostelium*, 49–51, 57–59, 80–84, 90, 91
  in myxomycetes, 114, 115, 135, 136
  in *Polysphondylium*, 57, 65, 66
  in protostelids, 16, 17, 31, 37
Stemonitida, 113–115, 136, 247
  in key, 108
Stemonitidae, 113, 135
Stemonitomycetidae, 114
Synaptonemal complexes
  in *Ceratiomyxa*, 31
  in dictyostelids, question of, 93
  in *Labyrinthula*, 236, 237
  in myxomycetes, 123, 140, 141, 147, 148
  in plasmodiophorids, 200, 209, 211
Synergism, 97

T

Temperature, effect of
  on migration in *Dictyostelium*, 78
  on mitosis in *Physarum*, 127

Tetrahymena, 49
Thraustochytriaceae, 216
Thraustochytriidae, 215, 228–230
  in key, 224
Thraustochytrids, 215, 228–230
  amoeboid protoplasts of, 229, 230
  aplanospores of, 230
  cell movement in, 221, 228
  distribution of, 218
  isolation and culture of, 218
  mitotic cell division in, 229
  phylogeny of, 250–252
  plasmodium-like protoplasts of, 229
  ultrastructure of, 237–241
Trichiida, 112, 113
  in key, 108
Trichiidae, 112

U

Ultrastructure
  of acrasids, 184–190
  of dictyostelids, 84–94
  of *Endemosarca*, 212–214
  of *Labyrinthula*, 230–237
  of myxomycetes, 141–152
  of plasmodiophorids, 206–212
  of protostelids, 38–43
  of thraustochytrids, 237–241

V

Variation in *Acrasis rosea*, 175
Virus
  in thraustochytrids, 241
  zoospores of *Polymyxa* as vectors of, 200

Z

Zoocysts of *Ceratiomyxella*, 21, 23–25
Zoosporangia
  of *Labyrinthula*, 223, 224, 237
  of *Phagomyxa*, 203
  of plasmodiophorids, 195–197
  of proteomyxans, 202, 203
  of thraustochytrids, 228–230, 239
Zoospores, *see* Flagellate cells

A 4
B 5
C 6
D 7
E 8
F 9
G 0
H 1
I 2
J 3